THE
EVERYTHING®
PARENT'S GUIDE TO
COMMON CORE MATH: GRADES K–5

Dear Reader,

The absolute best professional development I've ever attended focused on the Common Core State Standards for Mathematics and the Standards for Mathematical Practice, supported by the Massachusetts Department of Education. I am not kidding! For one or two weeks each summer for several years I had the opportunity to play with toys, draw with markers, create posters, and experience math like my students should. It was fun! It was also unlike the classes I took as a kid. I learned, among other things, that there is a huge difference between how a student *experiences* math versus having experiences *in* math. I always left these workshops with great ideas and anxious to get back to the classroom! I jumped at the opportunity to be involved with this book because I know there are many parents who are uncomfortable with the idea of their children learning math in a different way than they did. I've been in the same position, as a parent having my son ask me for help. I remember reading his workbook and then asking him, "Where's the rest?" because I saw no question and had no clue what he was being asked to do. So I hope this book brings you comfort with the Common Core and lets you see how your child will *experience* math.

Jim Brennan

Welcome to the EVERYTHING® Series!

These handy, accessible books give you all you need to tackle a difficult project, gain a new hobby, comprehend a fascinating topic, prepare for an exam, or even brush up on something you learned back in school but have since forgotten.

You can choose to read an Everything® book from cover to cover or just pick out the information you want from our four useful boxes: e-questions, e-facts, e-alerts, and e-ssentials. We give you everything you need to know on the subject, but throw in a lot of fun stuff along the way, too.

We now have more than 400 Everything® books in print, spanning such wide-ranging categories as weddings, pregnancy, cooking, music instruction, foreign language, crafts, pets, New Age, and so much more. When you're done reading them all, you can finally say you know Everything®!

PUBLISHER Karen Cooper

MANAGING EDITOR, EVERYTHING® SERIES Lisa Laing

COPY CHIEF Casey Ebert

ASSISTANT PRODUCTION EDITOR Alex Guarco

ACQUISITIONS EDITOR Lisa Laing

SENIOR DEVELOPMENT EDITOR Brett Palana-Shanahan

EVERYTHING® SERIES COVER DESIGNER Erin Alexander

Visit the entire Everything® series at www.everything.com

THE
EVERYTHING®
PARENT'S GUIDE TO
Common Core Math: Grades K–5

Understand the new math standards to
help your child learn and succeed

Jim Brennan

Avon, Massachusetts

To the educator heroes who lost their lives serving children: Rachel D'Avino, Dawn Lafferty Hochsprung, Victoria Soto, Mary Sherlach, Lauren Rousseau, and Anne Marie Murphy at Sandy Hook Elementary School, Newtown, CT, and Colleen Ritzer, Danvers High School, MA.

An Everything® Series Book.
Everything® and everything.com® are registered trademarks of F+W Media, Inc.

Published by
Adams Media, a division of F+W Media, Inc.
57 Littlefield Street, Avon, MA 02322. U.S.A.
www.adamsmedia.com

ISBN 10: 1-4405-8680-2
ISBN 13: 978-1-4405-8680-4
eISBN 10: 1-4405-8337-4
eISBN 13: 978-1-4405-8337-7

Printed in the United States of America.

10 9 8 7 6 5 4 3 2 1

Library of Congress Cataloging-in-Publication Data

Brennan, Jim,
    The everything parent's guide to common core math, K–5 / Jim Brennan.
        pages cm. -- (Everything series)
    Includes bibliographical references and index.
    ISBN 978-1-4405-8680-4 (pb : alk. paper) -- ISBN 1-4405-8680-2 (pb : alk. paper) -- ISBN 978-1-4405-8337-7 (ebook) -- ISBN 1-4405-8337-4 (ebook)
    1. Education--Parent participation. 2. Mathematics--Study and teaching (Early childhood) 3. Mathematics--Study and teaching (Elementary) I. Title. II. Title: Parent's guide to common core math, K–5. III. Title: Guide to common core math, K–5.
    QA135.6.B725 2015
    372.7--dc23

                                2014049035

Illustrations by Eric Andrews.

This book is available at quantity discounts for bulk purchases.
For information, please call 1-800-289-0963.

Contents

Acknowledgments

I would like to thank Lisa M. Laing, Managing Editor, Everything® series, for her patience, support, and excellent feedback, and Erin Dawson for her consultations on art, drawings, and math rendering. I would like to thank my daughter Julia for her editing and review expertise, my wife Lisa for her support and understanding, and my family and friends for their encouragement. I would also like to thank Patti Morsillo for being the model of how parents can support their child and their child's teacher.

Introduction

*No one will pay you for what you **know**, because Google knows everything!*
*People will pay you for what you can **do**!*

—Speaker at a conference addressing students and teachers

Today's students are preparing to enter a world in which colleges and businesses are demanding more than ever before. To ensure all students are ready for success after high school, the Common Core State Standards establish clear, consistent guidelines for what every student should know and be able to do in math and English language arts from kindergarten through 12th grade.

—Common Core State Standards Initiative

EVERY STATE IN THE United States has a department of education that sets policies, such as adopting standards for each of its school districts. The local school districts can be responsible for creating their own curriculum, or the districts may empower their own schools to create the curriculum. In Massachusetts, for example, each town has its own school board. Other states, such as Maryland, have county-based school boards that operate at a much larger scale, some with budgets in excess of $1 billion.

Over the last several years many states have agreed to adopt a common set of education standards, initially for English and Math, called the Common Core Standards. Even though your state may have adopted these standards, each state still operates independently. Curriculum and instructional decisions are still determined by local school districts and teachers. In other words, the standards tell schools *what* to teach, and the schools decide *how* they will teach it.

The Common Core State Standards are designed to identify the most essential skills and knowledge students need. The standards do not specify how students acquire these skills; that task is left to the local school districts and the teachers.

As a parent, the thought of a new standard of learning may sound worrisome, but the Common Core Standards are not as daunting as they may seem. The promise of the Common Core Standards is to bring more focus, more coherence, and more rigor to math education. That statement alone can be intimidating. It is important to understand each of these shifts in order to support students in their education.

- **Focus:** The standards call for greater focus in mathematics. What this means is that instead of racing to cover topics in today's mile-wide, inch-deep curriculum, teachers use their power to narrow and deepen the way time and energy is spent in the classroom. The standards require depth instead of breadth, students developing strong mathematical foundations with basic concepts and solid conceptual understanding, which means they understand why and how the math works. The standards also require that students really understand the procedures related to the skills, and can fluently use these skills to solve problems.
- **Coherence:** Coherence is all about the glue that holds everything together. It is about linking information from one grade to the next and building on students' prior knowledge. The standards have been designed around coherent progressions from grade to grade and in order to maintain coherence, it is important that a common language be used at each and every grade level. Coherence is also maintained through the ways in which the major topics are reinforced. Instead of teaching individual skills and topics in isolation, it is about embedding skills into grade-level word problems that allow for application and reinforcement, so students can make their own connections to the information and become stronger problem solvers.
- **Rigor:** The word *rigor* always seems to make things sound hard or more difficult to understand, but this is not the case. In the context of the Common Core Standards, rigor means that rather than having students learn how to calculate something and then move on to the next topic, the students have to *apply* what they learned. Rigor is creating an

environment in which each student is expected to learn at high levels, and each is supported so he or she can learn at high level, and each student demonstrates learning at high level.

The following example demonstrates three levels of rigor: conceptual understanding, procedural skill fluency, and application.

Sample Questions Addressing the Components of Rigor

Procedural Skill Fluency: Mark each equation true or false: $8 \times 9 = 80 - 8$, $7 \times 5 = 25$, $8 \times 3 = 46$

Conceptual Understanding: What is the value of $3x + y - 3(x + y)$ when $x = 18.22$ and $y = -1$?

Application: On Monday Billy walked $\frac{1}{2}$ mile. On Tuesday Billy walked some more. Altogether Billy walked $2\frac{1}{2}$ miles. How far did Billy walk on Tuesday?

Asking students to apply what they know to solve real-world problems further solidifies their overall understanding of mathematics. Students become active members of the learning environment, demonstrating their abilities through the mathematical tasks, and meeting the academic demands of the Common Core State Standards.

CHAPTER 1

The Common Core Standards for Mathematics

Standards-based education is not a passing fad. As your child enters prekindergarten, kindergarten, or anywhere from grade 1 to grade 5, you, your child, and your child's teacher will be heavily vested in the Common Core State Standards. States are transitioning from their state standards to their state's adaptation of the Common Core Standards. Some states have a longer-term plan than others, providing teachers, parents, and students time to adjust; other states or districts have taken a much more abrupt approach. This chapter presents the Common Core State Standards and their role in providing your child with the required twenty-first-century skills.

The Big *W*'s of the Common Core Standards

This section provides answers to the Big *W* questions of What the standards are, Why they are needed, Which states have adopted them, and What's different in the Common Core State Standards.

What Are the Common Core Standards?

Quoting the Math Common Core Coalition, the mission of the Common Core State Standards is to provide "a consistent, clear understanding of what students are expected to learn, so teachers and parents know what they need to do to help them" and ultimately prepare them for college and careers. The Common Core State Standards for Mathematics specify standards at each grade level, and are divided into certain domains such as Number and Operations in Base Ten, Measurement and Data, and Geometry. Groups of similar standards within each domain are grouped into clusters. The K–5 standards define what should be covered over the course of a school year.

QUESTION

What is the difference between *standards* and *curriculum*?
Standards, such as the Common Core Standards for Mathematics, describe what students should be able to do, and what they should know. *Curriculum* describes a sequence of units of study, with instructional goals, objectives, and measurable student outcomes. Teachers follow a curriculum to create lessons and activities to direct student learning to achieve outcomes. Standards are usually organized by topic; they do not define a sequence, nor do they provide teachers with lessons they can use in the classroom.

Why Do We Need the Common Core Standards?

The U.S. Department of Education provides results that show that while some individual states rank well when compared internationally, the United States is not among the top performing countries in mathematics. In fact, in the 2012 Programme for International Student Assessment (PISA) the United States ranked 25 out of 34 countries in mathematics.

State governors collectively worry that the United States is not preparing students for the changing world economy; they see other countries making measurable progress in math and science initiatives while the United States does not. Additionally, students moving between districts did not experience consistency, as the prerequisite grade-level skills varied between states. So if a child was to move from state to state, or even district to district within the same state, they could be either dramatically ahead or behind their new class. Assessment and achievement scores from one school were difficult to compare to assessment scores from another, particularly when the test items were not similar and there were measurable differences in what was being taught.

What States Have Adopted the Common Core Standards?

The states that have *not* adopted the Common Core Standards are Alaska, Texas, Nebraska, Minnesota, Indiana, and Virginia. Puerto Rico has also not adopted the Common Core Standards, but all other U.S. territories have. Several states are reviewing their commitment to the Common Core Standards and to the common assessments. For the latest information, see the website for the department of education in your state.

What's Different in the Common Core Standards?

The Common Core Standards are not a repackaging of previous standards. Of course, it's a hard sell to say "this isn't your parents' math," because then (rightfully so) you as a parent won't be comfortable with the changes. There are three major shifts within teaching mathematics that you'll need to understand in order to help your child succeed:

1. **Focus:** The standards make sure there are *fewer* big ideas being taught, thus the learning concepts that are being taught are emphasized.
2. **Coherence:** There is a progression of topics and performances that are developmentally connected to other progressions. Therefore each concept your child learns will be linked to another.
3. **Rigor:** Your child will be able to apply concepts and skills to new situations.

The Common Core State Standards are designed so that teachers (and students) no longer "dig a mile wide, yet only an inch deep," as the previous standards have been characterized as doing.

As a parent, you need to be open to the fact that how your child will be taught today is different from the way you were taught. As classrooms move from being teacher-centered to student-centered, informed parents are a tremendous asset in helping students to be more active and accountable in their own learning.

The Big *C*'s of the Common Core

The Partnership for 21st Century Skills (*www.p21.org*) has identified four components of twenty-first century skills as Critical Thinking, Communication, Collaboration, and Creativity. These represent skills that standardized assessments don't measure well, and they are critical for success in college and in a career. They are referred to here as the Big *C*'s.

The Common Core State Standards are designed to increase mathematical achievement in the United States, and prepare students for success in either college or a career. Students with a deep understanding of math, problem solving, and interpersonal skills will find themselves in high demand in science and technology, and reasoning skills apply to all fields.

Critical Thinking

Critical thinking is not restricted to the math domain. Why do students answer questions in history, or compare and contrast characters in the study of literature, but have to "solve problems" in math? Are math teachers the only ones giving your child *problems*? Critical thinking skills are required by students to collect, process, and present information. Students learn to critique, analyze, compare, contrast, and construct examples and counterexamples in math, as they do in all subjects.

Communication

Your child receives information from more sources than at any other time in history. Students use communication systems for long-distance learning, and before they graduate from high school they will be passing in their homework electronically, using collaborative resources to communicate with other students and their teachers. At each level, they are required to read and to write clearly, using specific and appropriate language.

Collaboration

Successful students work collaboratively with others. The work world they will encounter after completing high school or college will require thinking and working in groups. Students integrate communications, critical thinking, and creativity to produce work superior to what they can produce alone. The math standards lend themselves well to small-group work where students identify patterns, number relationships, compare answers, and, most importantly, learn the processes used by other students to solve problems.

Creativity

It has been said that the great thing about preschool and kindergarten is that the math standards can be delivered within the context of play, and that math development comes from the unstructured ways in which children naturally sort, compare, and count. Under the Common Core State Standards for math, by delaying the teaching of rote calculation algorithms, and allowing students to make multiple representations of a problem, students develop creative problem-solving skills. Students also develop modeling skills, and learn to appreciate that problems can be solved in more than one way. When they are finally taught the standard algorithms, students have a deeper understanding of them, and find creative ways to apply them.

CHAPTER 2

The Standards for Mathematical Practice

How your child *experiences* mathematics will be substantially different than how you experienced learning math. The Standards for Mathematical Practice are used by teachers as a guide to help them design lessons and activities to provide students with experiences that lead them to develop student skills as mathematicians. Of course, not everyone will become a mathematician, but everyone will use math throughout their lives. The eight practices are not tied to math content. You can think of them as an eight-car train that delivers math content. As you read through the practices, look for ways that these practices will prepare your child with twenty-first century learning skills covered in the previous chapter: Critical Thinking, Communication, Collaboration, and Creativity.

Why the Standards for Mathematical Practice Are Important

Think back to when you were in school. Were you ever given a worksheet with practice problems that you were able to answer without much effort? That type of activity doesn't really provide a student with a very rich mathematics experience. What kind of experience arises when a worksheet is given to a child who hasn't been too successful with math in the past, has a lot of math anxiety, and can't answer any of the questions? One of the teacher's many tasks is to create lessons that are challenging, maybe a little frustrating, but not so frustrating that a child disengages.

The math practices are intended to help students remain engaged with math and to let them know that it's okay to attempt to solve problems by trying different methods, including some that might not lead to answers. A few children develop "learned helplessness"; if they can't see how to start thinking through a problem, they'll just say "I don't get it" and wait for someone to rescue them. The child who gets rescued is shown how to do a problem by having someone else do all of the thinking for them. Maybe the student can then solve a similar problem with different numbers the same way, but learning and understanding is minimized instead of maximized.

The eight Standards for Mathematical Practice, as established by the Common Core State Standard Initiative, are as follows:

1. Make sense of problems and persevere in solving them
2. Reason abstractly and quantitatively
3. Construct viable arguments and critique the reasoning of others
4. Model with mathematics
5. Use appropriate tools strategically
6. Attend to precision
7. Look for and make use of structure
8. Look for and express regularity in repeated reasoning

In this chapter, you'll explore each of the eight Standards for Mathematical Practice, first as defined by the 2014 Common Core State Standards Initiative, and then through a more user-friendly discussion about key points within the standard. Each section describes what the student should experience, and provides ways that you can help develop that experience. Rather than *showing* your child a process that can be followed to find a particular answer, *guide* him through the process by asking him questions: questions about what he sees, what he thinks a problem is asking, or why he chose a particular strategy. Students learn from what they see, hear, and read, but they will learn *more* from what they do and what they say.

Make Sense of Problems and Persevere in Solving Them

Mathematically proficient students start by explaining to themselves the meaning of a problem and looking for entry points to its solution. They analyze givens, constraints, relationships, and goals. They make conjectures about the form and meaning of the solution, and plan a solution pathway rather than simply jumping into a solution attempt. They consider analogous problems, and try special cases and simpler forms of the original problem in order to gain insight into its solution. They monitor and evaluate their progress and change course if necessary. Older students might, depending on the context of the problem, transform algebraic expressions or change the viewing window on their graphing calculator to get the information they need. Mathematically proficient students can explain correspondences between equations, verbal descriptions, tables, and graphs or draw diagrams of important features and relationships, graph data, and search for regularity or trends. Younger students might rely on using concrete objects or pictures to help conceptualize and solve a problem. Mathematically proficient students check their answers to problems using a different method, and they continually ask themselves, "Does this make sense?" They can understand the approaches of others to solving complex problems and identify correspondences between different approaches.

Identify Entry Points to a Solution

Students start understanding a problem by explaining to themselves what a problem is asking, what relevant information is being provided, and identifying entry points toward its solution. Some teachers will prepare templates for students to use to help them identify the parts of a word problem. What is the question asking? What information do I need to answer the question? What information is given in the question? What information can I find out, using the information that I have?

Helping Your Child Find an Entry Point

You can help your child understand the problem by asking him, "What is the meaning of the problem?" or "Have you seen similar problems?" You can help him identify the parts that he *needs* and the ones he *has*. Young learners may not understand all of the vocabulary in a word problem, and they may have more questions about the reading aspect of a problem than the math aspect.

Examples from K–5

Suppose your child is trying to find the area of a rectangle. Is he clear about the difference between area and perimeter? Ask him what measurements he needs to find the area, and ask if those measurements are given. Sometimes one of the dimensions must be calculated first.

Plan a Solution Pathway

Different students answering the same question may find many different ways to solve the problem; they have different *pathways* to the solution. Students will make a plan for solving problems, possibly thinking about what the end product should be and finding examples and non-examples of similar problems. Students will ask themselves what successful examples have in common. To help in understanding the question and the format of the answer, your child may decide that a visual model or equation would help find the solution.

Helping Your Child Plan a Solution Pathway

Ask your child specific questions about why she has chosen to use a particular operation or selected a certain visual model. Make sure you don't frame the question in such a way as to suggest that it was a good or bad choice. As your child explains why she choose particular methods, she further develops her communication skills. If your child understands the problem but is having trouble creating a plan, you can give her a hint in the form of a multiple choice question such as, "Do you think it would be easier to figure out the answer using a number line or an equation?"

Examples from K–5

Students in early grades are asked to classify shapes. For example, they may be asked to classify a trapezoid. It may help them first to see that it is not a circle, because it doesn't have any curves, and it isn't a triangle because it has more than three sides. One approach a child may use would be to first pre-sort (or categorize) shapes they can identify to help simplify the problem.

Monitor Progress, Change Course If Needed

When students have understood what a question is asking they can examine their work in progress and ask themselves if they think they are closer to a solution. Sometimes after working on a problem for a while a student might discover that the path they are on will not lead to a solution. For example, they may have used a graphical representation that doesn't present data in a format that will help them answer questions about the data. After examining their work they may discover their diagram does not represent all of the information that was provided in the question and they are able to select a better format. Students need to develop the skills to pause and ask themselves if they think they are still on a path to a solution. When students are answering a multistep problem, they can stop in the middle and verify that the calculations they have made so far are correct to enable them to continue with confidence; they may also discover they had made calculation errors or change course based on problem solving decisions they have made.

Helping Your Child Monitor Progress and Persevere

Like many children, your child may see arriving at an answer as being finished with the problem. Helping your child learn early in her education that checking her answer not only improves her problem-solving skills, it also provides her with an opportunity to make any corrections that are needed. For example, when she solves a multiplication problem, verifying her answer may require division. In this way your child gains a deeper understanding of how the operations are related. A good question to always ask your child is, "How do you know when you are finished? "This prompt will help your child think about the answer she is trying to find and to put the answer into the context of the question. Is she multiplying to find a total quantity of something? Or the square area of a room?

Reason Abstractly and Quantitatively

Mathematically proficient students make sense of the quantities and their relationships in problem situations. Students bring two complementary abilities to bear on problems involving quantitative relationships: the ability to decontextualize—to abstract a given situation and represent it symbolically, and manipulate the representing symbols as if they have a life of their own, without necessarily attending to their referents—and the ability to contextualize, to pause as needed during the manipulation process in order to probe into the referents for the symbols involved. Quantitative reasoning entails habits of creating a coherent representation of the problem at hand; considering the units involved; attending to the meanings of quantities, not just how to compute them; and knowing and flexibly using different properties of operations and objects.

Make Sense of the Quantities and Their Relationships in Problem Situations

Students will use a variety of visual models or manipulatives to represent quantities, relationships between quantities, and the underlying mathematical structure of a problem situation. Math manipulatives can be everyday objects that can be counted, sorted, or used to model mathematical concepts. They can be math specialty products such as pattern blocks or base 10 blocks.

Helping Your Child Relate Numbers

When your child has problems that involve having some number of objects and getting more, there is a relationship between how many he started with and how many he ended up with. Early learners need practice in order to understand the relationships between the numbers involved. For example, if the number added is greater than zero, then the number they end with will always be greater than the number they started with.

You can help your child understand the quantities by asking him to describe the relationships he sees. Ask him what he would expect to happen if he has 20 objects and added more, or took some away. When using graphs or plots, you can help your child interpret data by asking him about what he sees, about the most (or least) of something being represented, or about the direction of a line.

Examples from K–5

A student is asked to find the perimeter of a rectangle. The length of the rectangle is 2 feet, and the width is 5 inches. The student draws a rectangle, labels all sides 2, 5, 2, and 5, and adds them together to find the perimeter is 14 . . . but what are the units he assigns? The same units need to be used for all lengths. So he goes back to the basic data provided and reinterprets it. The 2 feet can be converted to 24 inches, and the perimeter becomes $24 + 5 + 24 + 5 = 58$ inches.

Decontextualize

Students may use manipulatives or numerals to represent the quantities of objects in a word problem, such as, "John has 3 apples and buys 2 more." This can be modeled by an equation $(3 + 2 = 5)$ or represented this way:

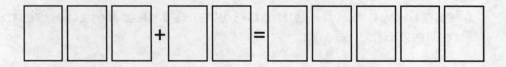

This takes away the context that the student is talking about apples; the apples are represented by symbols, and the symbols are manipulated to solve the problem.

Helping Your Child Decontextualize

In grades K–5, your child will be using visual models and equations. The *best* way to represent a problem isn't always apparent, but some visual models may be better than others. You can help by asking questions about which model might be better to use for a particular problem. Consider whether the problem is using whole numbers or fractions, or perhaps the operations may lead your child to decide on one representation being better than another.

Examples from K–5

When students are given word problems that involve real objects (such as having two dozen tennis balls and three dozen golf balls) and are asked "How many are there altogether?" or "If they give one golf ball away, how many more golf balls will there be than tennis balls?" they *decontextualize* when they represent the objects with numbers (or symbols). When students write an equation to find the total $24+36=?$, they have decontextualized the problem. If a third person were to read only the numeric representation, there is no context whatsoever that identifies these numbers as tennis balls or golf balls.

Contextualize

During a problem-solving process, students may have represented quantities of some objects with numbers. Both during the problem-solving process and after arriving at a numeric answer, students need to identify what the numbers represent. For example, during a calculation about walking certain distances and representing those distances on a number line, if a position ends to the left of 0 on a number line (indicating a negative integer), your child should stop to see if that number makes sense for the problem. "If I walked 3 miles on one day, and 10 more another day, how could I walk less than 0 total miles?"

Helping Your Child Contextualize

At any point during a problem-solving process, you can check your child's understanding by asking "What does your letter t represent?" or "I see your answer is 8, but what does the number 8 represent? Eight of what?"

Examples from K–5

Consider the following problem:

> Mrs. Sutton, the gym teacher, has 24 tennis balls and 36 golf balls. She loans 8 golf balls to Ms. Simmons. How many more golf balls does Mrs. Sutton have than tennis balls?

After loaning 8 golf balls, Mrs. Sutton has $36-8=28$ golf balls left. When students set up the answer as a final equation, they write $28-24=4$. What does the 4 represent? Students *contextualize* when they put the 4 back into the context of the problem, Mrs. Sutton has four more golf balls than tennis balls.

Construct Viable Arguments and Critique the Reasoning of Others

Mathematically proficient students understand and use stated assumptions, definitions, and previously established results in constructing arguments. They make conjectures and build a logical progression of statements to explore the truth of their conjectures. They are able to analyze situations by breaking them into cases, and can recognize and use counterexamples. They justify their conclusions, communicate them to others, and respond to the arguments of others. They reason inductively about data, making plausible arguments that take into account the context from which the data arose. Mathematically proficient students are also able to compare the effectiveness of two plausible arguments, distinguish correct logic or reasoning from that which is flawed, and—if there is a flaw in an argument—explain what it is. Elementary students can construct arguments using concrete referents such as objects, drawings, diagrams, and actions. Such arguments can make sense and be correct, even though they are not generalized or made formal until later grades. Later, students learn to determine domains to which an argument applies. Students at all grades can listen or read the arguments of others, decide whether they make sense, and ask useful questions to clarify or improve the arguments.

Math Speak

Collaboration and communication are important twenty-first century proficiencies, and being able to communicate using math vocabulary is a critical skill. *Math speak* is a playful term to remind students to use specific vocabulary when they discuss a problem or a solution, and to look for specific vocabulary from others. Students learn to build arguments and present a sequence of logic and reason statements using that vocabulary. When children work with partners and small groups, teachers look for discourse that applies math vocabulary.

Helping Your Child with Math Vocabulary

You can help your child develop her reasoning skills with simple prompts to get her to say more and to use math vocabulary, such as "What did you mean by _____?", "Can you find another way to say that?", or "Can you say more about that?"

Examples from K–5

Aaron and Holly were working on the following homework problem:

Evaluate the expression $35 - 5 \times 5 - 2$.

Aaron said the answer is 148. Holly said the answer is 90. Aaron said he was clearly right, because $35 - 5 = 30$, $30 \times 5 = 150$, and $150 - 2 = 148$. Holly said she was clearly right because $35 - 5 = 30$, and $5 - 2 = 3$, and $30 \times 3 = 90$. Who is correct? Why?

Here your child needs to follow the reasoning of two students, agreeing or disagreeing with each step. He can't get away with just saying who is correct; he has to explain why. In this example, neither student had the correct answer because neither followed the order of operations, which stipulates that multiplication is done before addition and subtraction. The problem should have been interpreted as $35 - (5 \times 5) - 2$; where the multiplication is done first: $5 \times 5 = 25$, $35 - 25 = 10$, $10 - 2 = 8$. Aaron evaluated the expression doing the operations from left to right. Holly explained her answer as if the expression was $(35 - 5) \times (5 - 2)$, which does not follow the correct order of operations. The math speak in this example is *order of operations*, *subtraction*, and *multiplication*. This example also critiques the reasoning of others.

Connecting with Prior Experiences

Students build on what they know—vocabulary, processes, mathematical properties—to extend their thinking by making conjectures and exploring them logically.

Helping Your Child Use Prior Knowledge

You are able to make connections within your child's learning more fluently than your child can. You can help her to leverage what she knows and what she has done before to help her with what she is doing. If your child is struggling with a problem, you can jog her memory by talking about a problem she has solved before. Maybe the prior problem is similar, or maybe it acts as a counterexample of what she did before; even recalling previous mistakes can help set her on a solution path.

Listen and Critique

By viewing the work of other students or working together, students become more aware that there is more than one way to model and solve problems. When students don't arrive at the same answer using different strategies, they listen to the reasoning of others and they learn to ask questions to clarify that reasoning. Students develop communication skills when they explain their reasoning to others, and improve reasoning skills by answering questions to clarify their own explanations. Your child will also use drawings and physical models to help him construct arguments or counterexamples.

Helping Your Child Listen and Critique

As in the previous example, your child will encounter questions about the solutions found by other students. You can help your child build critiquing skills by posing "what if" questions, either while helping her with her homework or in routines in daily life. For example: "It's 8:30 now, and the dishwasher takes 40 minutes to wash the dishes. If I start it now, will you be able to unload it at 9:00?"

Model with Mathematics

Mathematically proficient students can apply the mathematics they know to solve problems arising in everyday life, society, and the workplace. In early grades, this might be as simple as writing an addition equation to describe a situation. In middle grades, a student might apply proportional reasoning to plan a school event or analyze a problem in the community. By high school, a student might use geometry to solve a design problem or use a function to describe how one quantity of interest depends on another. Mathematically proficient students who can apply what they know are comfortable making assumptions and approximations to simplify a complicated situation, realizing that these may need revision later. They are able to identify important quantities in a practical situation and map their relationships using such tools as diagrams, two-way tables, graphs, flowcharts, and formulas. They can analyze those relationships mathematically to draw conclusions. They routinely interpret their mathematical results in the context of the situation and reflect on whether the results make sense, possibly improving the model if it has not served its purpose.

Select a Model That Interacts with the Underlying Math

Students use models to represent quantities and operations. They begin by applying what they uncover by understanding a problem, and select a model to represent the problem. Then they generate a pathway toward the solution. Your child will experience many types of models and will discover how best to use them.

Helping Your Child Select a Model

Your child will be looking for a model that fits the problem. Number lines and tape diagrams are great models for addition, subtraction, and comparison problems. An area model is a good model for multiplication and division; collections of similar objects can be represented with repeated use of any easily drawn shape or common objects around the house. You can reinforce or guide your child's choice of model by asking him how he will use it. Examples of all these techniques are provided throughout Chapters 3 through 9.

Simplify Complexity

Students look for ways to simplify problems so that they can be managed by models. For example, to model the problem 52×75 students can use an area diagram; they can represent the 52 as $50 + 2$ and the 75 as $70 + 5$. Then $70 + 5$ can be represented with two columns, and $50 + 2$ can be represented with two rows. Finding the area of each box, and finding the sum of each box has simplified the process of finding the product of 52 and 75.

What Are Models?

Models can take many forms, such as tables, graphs, diagrams, and equations. They can be paper-based, or physical representations for objects, numbers, math problems, or operations. Your child will use different kinds of graphic organizers in all her classes to organize ideas, modeling a collection or partitioning of similar or dissimilar ideas and concepts.

Helping Your Child Understand Models

You can help your child make connections between different representations for the same problem; for example, connecting equations to area diagrams. You can help reinforce that modeling a problem will help him to understand it, and illustrate that models are a set of tools to help strengthen his problem-solving abilities. They are a fun way to represent the problems. Examples of models are provided throughout Chapters 3 through 9.

Use Appropriate Tools Strategically

Mathematically proficient students consider the available tools when solving a mathematical problem. These tools might include pencil and paper, concrete models, a ruler, a protractor, a calculator, a spreadsheet, a computer algebra system, a statistical package, or dynamic geometry software. Proficient students are sufficiently familiar with tools appropriate for their grade or course to make sound decisions about when each of these tools might be helpful, recognizing both the insight to be gained and their limitations. For example, mathematically proficient high school students analyze graphs of functions and solutions generated using a graphing calculator. They detect possible errors by strategically using estimation and other mathematical knowledge. When making mathematical models, they know that technology can enable them to visualize the results of varying assumptions, explore consequences, and compare predictions with data. Mathematically proficient students at various grade levels are able to identify relevant external mathematical resources, such as digital content located on a website, and use them to pose or solve problems. They are able to use technological tools to explore and deepen their understanding of concepts.

The Tools

The tools that are used strategically do not have to be electronic: Graph paper and a pencil can be tools, if used intentionally. Completing a homework assignment on graph paper because a student happens to have some around does not exercise this mathematical practice. Using graph paper to complete a homework assignment that is used strategically to keep multiplication problems neatly organized, however, *does* exercise this practice. When evaluating this practice, focus on the word *strategy*.

Use Estimation Strategically

Estimation is not only used when a problem asks for an estimate, or to answer an "about how many" question, it is an important tool for students to monitor their progress on a solution path. Rounding is another tool. Both estimation and rounding should be considered strategies your child can use as she would any other tool.

Helping Your Child Estimate Strategically

When your child is in the middle of a multistep problem and is uncertain if she is on the right path, have her make a quick estimate of what she has done so far. This will help her to choose between continuing down the same solution path, or backing up and making different decisions. In some cases, the problem itself does not specify that the answer should be an estimate, but sometimes there isn't enough information to provide an exact answer. You may need to help your child to identify problems that are implicitly asking for an estimate.

Select and Apply Technology Tools

Students can choose among tools that will help them identify a solution path, and can use technological tools to explore and deepen their understanding of concepts. This doesn't mean selecting a tool that will do the work for them; the focus should be on applying technology as part of the problem-solving process.

Helping Your Child Select and Use Technology

Even young children learn to use technology, and many elementary school children will have apps on smartphones or tablets that can create multiple representations of the data students can collect. In addition to selecting a representation for a problem, students can choose an appropriate technological tool to use. Spreadsheets or math software (such as

GeoGebra) can be appropriate tools. You can help your child by helping her select tools and understand when certain tools are appropriate or not appropriate. Don't be afraid to learn to use new tools yourself!

Helping Your Child Online

Your child may be good at using the Internet and conducting searches, but she needs help to identify appropriate, relevant, accurate, and safe online resources to further develop her mathematical proficiency. There are online tools you can help her find that are available for problem solving, websites that can provide practice problems, practice assessments, and online tutorials targeted to young learners; ask her teacher for input. Real-world connections can be made for careers that use math or supply real-life data.

Attend to Precision

Mathematically proficient students try to communicate precisely to others. They try to use clear definitions in discussion with others and in their own reasoning. They state the meaning of the symbols they choose, including using the equal sign consistently and appropriately. They are careful about specifying units of measure, and labeling axes to clarify the correspondence with quantities in a problem. They calculate accurately and efficiently, and express numerical answers with a degree of precision appropriate for the problem context. In the elementary grades, students give carefully formulated explanations to each other. By the time they reach high school they have learned to examine claims and make explicit use of definitions.

Use Specific Language in Communication

Mathematical statements are clear and unambiguous. It is clear what is known and what is not known. Students critique their use of words and the implied definitions of vocabulary used by others; they seek counterexamples and contradictions to deepen their understanding precise use of mathematics vocabulary and symbols.

Helping Your Child Build Vocabulary

Early learners are constantly building their vocabulary. You can help your child build vocabulary by using it yourself around the house. For example, refer to objects by their shape or other attributes, and use comparison language such as *similar*, *dissimilar*, *greater than*, *less than*, and *equals*.

Writing

One of the big *C*'s of the twenty-first century skills is Communication. Students need to write and explain their answers thoroughly using exact language. This goes beyond just making sure an answer to an area problem specifies the units. Students need to explain their steps and their understanding. "I did *this* because I calculated *value*, therefore my answer is *accurate answer*." Students can use similar template statements to learn how to write with enough detail to convey their understanding of the problem, the process they used along a solution path, and their answers or conclusions.

Helping Your Child Write for Math

You can help your child by evaluating her writing for accuracy, organization, clarity, and using the age-appropriate math vocabulary. In later grades, you can evaluate it for insight and mechanics. Stay connected with the math unit that your child is learning and see if she can correctly use the current vocabulary. You can provide suggestions about improving penmanship and helping her write complete sentences where appropriate.

Calculate Accurately

Students perform operations correctly, apply rounding appropriately, analyze the units for their answers, and label tables, graphs, and charts accurately and completely. Precision is reflected in both oral and written work.

Helping Your Child Calculate Accurately

You can check your child's work to make sure she has answered questions in the correct form (e.g., rounding answers to the number of places as directed) and using the correct units. If her answers seem off from the expected outcome, you can help her understand when to apply rounding at the appropriate time. Answering a question accurately also means she followed the order of operations and answered the question that was asked.

Examples from K–5

When students are adding two three-digit numbers, they sometimes forget to regroup (carry) into the hundreds or tens place, or they will use the regrouping action and lose one of the addends. Students that make an error in the first step of a multistep equation are more likely to provide an incorrect answer. The errors that students are most likely to find are with the last calculations they perform. That is why students should always check the "reasonableness" of their answers, and check their work starting from the beginning of the problem.

Look for and Make Use of Structure

Mathematically proficient students look closely to discern a pattern or structure. Young students, for example, might notice that three and seven more is the same amount as seven and three more, or they may sort a collection of shapes according to how many sides the shapes have. Later, students will see 7×8 equals the well-remembered $7 \times 5 + 7 \times 3$, in preparation for learning about the distributive property. In the expression $x^2 + 9x + 14$, older students can see the 14 as 2×7 and the 9 as $2 + 7$. They recognize the significance of an existing line in a geometric figure and can use the strategy of drawing an auxiliary line for solving problems. They also can step back for an overview and shift perspective. They can see complicated things, such as some algebraic expressions, as single objects or as being composed of several objects. For example, they can see $5 - 3(x - y)^2$ as 5 minus a positive number times a square, and use that to realize that its value cannot be more than 5 for any real numbers x and y.

Consider Behavior of Calculations

Your child will notice consistencies and relationships among objects, the results of calculations, and how calculations behave, especially in the early grades. When he begins to multiply by 5, he will see that adding another

5 results with a number ending either with a 5 or a 0. He will see patterns when adding 10 or 100 to a number, or multiplies a number by 10.

Helping Your Child Observe Calculations

Every time that zero is multiplied by a number, the answer is 0. When 1 is multiplied by some quantity (q) the answer is always q. Your child will notice countless patterns in the multiplication table or an addition table, or in the way that the house numbers are increasing by 2 (or 20) on one side of the road. You can help your child by asking him to identify patterns in day-to-day activities. If you are helping him with his homework and you see a list of numbers, ask him if he sees a pattern in the numbers. Your questions can enrich the homework experience; if he replies that his teacher didn't ask them to find patterns, just get the answer, then you can reply, "Yeah, but it's fun to find patterns, and it will help you with your homework next week."

Examples from K–5

Students learning to add three single-digit numbers, such as $3 + 5 + 7$, will "find the tens" first by adding $3 + 7 = 10$, then calculating $10 + 5 = 15$. They use the Commutative Property of Addition to rewrite $3 + 5 + 7$ as $3 + 7 + 5$, and then they apply what they know about adding 10 to a single digit.

Surface Underlying Structure

Students will use properties and rules of operations to uncover form and structure. For example, some children find adding two numbers together is easier when the higher number is presented first, and given a problem such as $3 + 7$, they may rewrite it as $7 + 3$. Students will find creative methods for finding patterns and using patterns to help them solve problems.

Helping Your Child Reveal Structure

Your child will work often with place value; you may find opportunities to help your child use the expanded form of a number, such as $322 = 300 + 20 + 2$ to help uncover the underlying structure of a number, help perform a calculation, or help your child compare numbers. Sometimes delaying a calculation, even when an expression seems more complicated than needed, can help to surface some hidden meaning.

Examples from K–5

The swim team bought 200 packages of popcorn for $2.00 per bag for a fundraising activity. They sold each package for $3.00. How much profit did they make? Intuitively, you may want to find the total cost of the popcorn and subtract it from the amount raised. Students can see the structure of the problem differently by delaying the calculation:

$$3.00 \times 200 - 2.00 \times 200 = 1.00 \times 200$$

Look for and Express Regularity in Repeated Reasoning

Mathematically proficient students notice if calculations are repeated, and look both for general methods and for shortcuts. Upper elementary students might notice when dividing 25 by 11 that they are repeating the same calculations over and over again, and conclude they have a repeating decimal. By paying attention to the calculation of slope as they repeatedly check whether points are on the line through (1, 2) with slope 3, middle school students might abstract the equation $\frac{(y-2)}{(x-1)} = 3$. Noticing the regularity in the way terms cancel when expanding $(x-1)(x+1)$, $(x-1)(x^2+x+1)$, and $(x-1)(x^3+x2+x+1)$ might lead them to the general formula for the sum of a geometric series. As they work to solve a problem, mathematically proficient students maintain oversight of the process, while attending to the details. They continually evaluate the reasonableness of their intermediate results.

What This Means for Your Child

You are probably familiar with the "guess and check" strategy, where the idea is to read a problem, guess an answer, then rework the problem to see if you are right. The process is repeated, guessing numbers each time that get you closer to the correct answer, until you get the correct answer itself. Here's an example:

Joey counts the money he has been saving. He has $98. Joey uses a guess and check strategy to see how many movie tickets he can buy, if each ticket costs $7. He guesses 10 tickets, and calculates that 10 tickets cost $70, which leaves him with $28. He knows 28 > 7, so he can buy more tickets. So he guesses a higher number. He guesses 15 tickets, and calculates that 15 tickets is $105. That is more than what he has, so he guesses lower. Joey then guesses 12 tickets, and calculates that 12 tickets cost $84, which leaves him with $14. He notices that 14 is exactly enough for 2 more tickets, so he guesses 14 tickets, and verifies that 14 tickets cost $98.

The guess and check strategy is tried and true, but a slightly different strategy called "guess, check, and generalize" will help students gain insight into the structure of numbers and give them a more general-purpose tool. Students are first told to select three random numbers and use these numbers to see if they are solutions to the same word problem. They keep track of calculations in a table.

Joey counts the money he has been saving. He has $98. Joey uses a guess, check, and generalize strategy to see how many movie tickets he can buy if each ticket costs $7.

Joey tests 10, 3, and 5 tickets:

10 tickets cost $70, $98 - 70 = 28$ there is money left over

3 tickets cost $21, $98 - 21 = 77$ there is money left over

5 tickets cost $35, $98 - 35 = 63$ there is money left over

To generalize, Joey uses t for the number of tickets:

t tickets cost $\$7t$. $98 - 7t = 0$, because Joey wants to use all of his money. So a general strategy is for Joey to find the missing factor of 7, that makes $98 - 7 \times \underline{\hspace{1cm}} = 0$ true.

By using the guess, check, and generalize method Joey will see the structure of the problem through repeated reasoning. He will be able to make generalizations about the problem and apply his process to solve similar problems. To reuse this for $119, he would use a factor of 7 that makes $119 - 7 \times \underline{\hspace{1cm}} = 0$ true.

Your child will experience repeated reasoning when they are physically modeling problems with coins or other manipulatives, solving puzzles, or constructing (or deconstructing) with blocks or similar materials.

Look for Repetition in Calculations

Students will notice when the same calculation is being done repeatedly; they may see a "rhythm" in the operations (or in physically making the same motions). The repetition in calculations may help students to find short cuts. The guess-check-generalize process is very helpful to identify repetition in calculations. When students see repetition they can find new ways to solve problems, and in later grades, generalize calculations using variables.

Help your child keep his work organized, so if there are repeating calculations he is more likely to see it.

Look for General Methods and Short Cuts

Your child may be asked to answer "what's my rule" questions that show a pattern, or provide a table for the student to complete. These types of exercise build the experience of finding patterns as well as developing number sense. For example, find the missing numbers, and explain your reasoning: 2, 6, 10, _____, 18, 22, _____.

Helping Your Child Look for Shortcuts

Your child may notice that the same two calculations are being done back to back, but might not know how to describe it. You can help him describe it by adding a letter to represent something that is common between calculations; that may help him make generalizations. You may be able to ask him questions that will allow him to see how two separate calculations can be done together. For example, when he is finding the perimeter of a square, rather than adding the four sides together using three addition calculations encourage him to use one multiplication operation, 4 times the side length.

Maintain Oversight of the Process While Simultaneously Attending to Details

Your child may get projects or multi-part questions that require multiple steps. Problems like these call for considerable management that needs ongoing attention. Keeping his work organized and adding headings or labels can help keep the details clear to him. During a long, multistep process, your child can pause periodically to ask himself "Where am I in the process?" and "Have I skipped any steps?"

Helping Your Child Maintain Oversight of the Process

When your child begins to answer word problems and begins to decontextualize something, he can lose track of what the numbers represent. For example, there may be 9 cars in the problem, and then the number 9 is used throughout the problem-solving process. When there are many objects that are added or subtracted, it is easy to lose track of what the calculations mean. You can help him oversee the process by asking questions like, "What does the 180 mean? Is it the number of cars, or the number of miles?"

Helping Your Child Evaluate Intermediate Results

With multistep calculation, the result of one step is carried forward for a calculation into the next step. Prior to using the answer in the next step, ask your child if the intermediate number makes sense, or is it close to what he was expecting. Estimation and rounding are good strategies to use to help evaluate how reasonable an answer might be. Between steps, ask your child what he is expecting. Putting a number back into the context of a problem can help answer this question. For example, if he is adding 10 cars to 14 red cars to see if they will fit in a box that can hold 30 cars, and his intermediate answer representing the total number of cars is 4, ask him if he starts with 10 cars and adds more cars, can he have less than 10?

CHAPTER 3

Prekindergarten

While the Common Core Standards are defined for kindergarten through grade 12, the majority of states have also defined learning goals for prekindergarten children. Children might attend preschool or pre-K programs for multiple years leading up to kindergarten, and there are differences in math skills between a three year old and a five year old. The skills identified in this chapter apply to children preparing to enter kindergarten the following year. Checklists provided at the end of each section of this chapter can help you measure the mathematical growth of your child throughout the years preceding kindergarten.

What Your Prekindergartener Is Expected to Learn

Pre-K students are not expected to experience direct math instruction from a teacher, but instead acquire their math skills through play and everyday activities. Pre-K teachers can be very clever at integrating math into ordinary tasks and instilling a sense of pride into their students regarding their mathematical accomplishments.

Pre-K students will learn to count, focusing on numbers from 1 to 10, and they will want to count everything! As they count out loud the number of items in a group and stop when all of the items have been counted, they will understand that the last number they said represents the number of items in the group. They will learn the names of each of the numbers, be able to say them, and identify them when they are represented numerically. They will be able to match a group of items with the word that represents the size of the group and its numeric representation. They will understand addition as adding things to the group, and subtraction as taking away things from the group.

ESSENTIAL

Games, such as the card game UNO, are a wonderful way to get children familiar with numbers. Appropriate cards can be selected from the large UNO deck, and kids play without performing calculations. They develop number and color matching skills, along with a sense of direction. UNO is available in many different themed sets, making it a very popular game with kids of all ages.

Pre-K students also learn measurement and geometry by observing and interacting with the physical world around them. Your child will learn about shapes like circles and squares. He'll compare measurements and wonder "Who's taller?" or "Who's older?" He'll learn how to express spatial relationships, such as "It's behind the chair," and he'll sort and categorize objects by their attributes. Kids love to sort stuff; they'll sort anything. Experiment by placing random objects in a bag and handing it to a child without giving him any instructions; he will take the items out of the bag and sort them. He will determine for himself what attributes will be used for sorting. See if you can figure out how he has sorted them.

Prekindergarten Mathematics Standards

There are four general categories of math skills your child will be learning in prekindergarten, and certain abilities that fall under each of those categories:

COUNTING
- Know number names and the counting sequence
- Count to tell the number of objects
- Compare numbers

ADDITION AND SUBTRACTION
- Understand addition as putting together and adding to, and understand subtraction as taking apart and taking from

MEASUREMENT
- Describe and compare measurable attributes

GEOMETRY
- Identify and describe shapes (squares, circles, triangles, rectangles)
- Analyze, compare, create, and compose shapes

At early ages, math skills are best assessed via observation or from a close examination of drawings/illustrations/constructions created by your child. Teachers have specialized training in looking at student work, but often asking a child how she is thinking about a problem or completing a task will help you understand her math reasoning, counting, or computational skills.

Counting

Pre-K math is all about whole numbers. This means no fractions, negative numbers, or decimals. Children may encounter these non-whole numbers in their daily lives—for example, outside temperature (less than 0), needing $1.25 to get a drink from a vending machine, or eating half a sandwich. But pre-K focuses on things kids can count, especially things they can pick up, move around, and play with. At the end of the pre-K year, your child will be expected to be able to count at least to 10. Of course it's fine to learn to count to 10 by first learning to count to 5. Even at the earliest ages, a good problem-solving strategy is to first solve a simpler problem.

FACT

A whole number that tells how many objects are in a group is called a *cardinal number*. Children use the cardinal numbers, but not the actual term *cardinal number*. You may encounter this term often in material about math content in the early grades.

Pre-K Counting Vocabulary

Children learn an amazing amount of vocabulary in pre-K. There is a critical need for young children to see, hear, write, and read as many words as possible. Pre-K children gain vocabulary by interacting with their environment, through the course of play, competition, games, and activities in school and at home. The amount of learning through experience actually exceeds the amount acquired through direct teaching! Children explore, and they figure things out.

The vocabulary for counting begins with the numbers one, two, three, four, five, six, seven, eight, nine, and ten, and then zero. Your child will be asked to count, order, and sort. She will perform counting activities, answer "how many" questions, and refer to objects in a sequence as being first, last, or in the middle.

Number Names and the Counting Sequence

In pre-K, your child will learn to count the number of items in a collection. Counting up to ten items is a good target, and some children will be able to count to twenty or higher. Your child will learn to recognize the numeral that represents each quantity. For example, she will see a collection of five pencils and know that this quantity can be represented with the number 5. Conversely, when she sees the number 5 she should be able to represent this with five items, like holding up five fingers. Your child will begin to write the numerals up to 10.

EXAMPLE 1

In this "count match" activity, have your child count the number of items within each square, and have him draw a line to the numeral that represents that quantity. You can discuss what a "line" is and can provide a ruler or a straight-edge if you would like to see neat straight lines. Without tools, you may see some very creative ways for drawing a line between a number and its corresponding image.

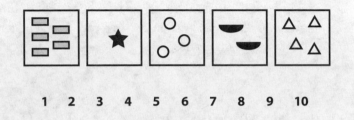

1 2 3 4 5 6 7 8 9 10

Discussion

As pre-K children apply what they know about counting and numbers to complete tasks like these, they demonstrate that they can count the number of items, in this case up to six, and they can match the number they count with the numeral that represents the number of items they counted. The numbers at the bottom are listed in order. This provides an additional way for a student to connect the number of items on the card with the numbers. Maybe after counting six squares on the first card a student begins with the number 1 at the bottom, and counts over to the sixth number. The numbers being in order provide a support system for the student to answer the question. Students who use this method are recognizing *correspondence*. Students will use this method until they are accustomed to recognizing the numbers.

EXAMPLE 2

Play the Show Me How game with your child. You pick a number and ask her to show you two ways to represent the quantity. Vary the questions as you play. Ask her to show you ways to show 5, then 7. Ask for more than two ways, or add to the game element by asking her to show you how many ways she can represent the number 8 in eight minutes.

Discussion

You can, for example, start with 5. Maybe she will write the number 5, maybe she will draw five of the same object, or maybe she will show you five fingers. By requiring children, even at early ages, to give more than one representation of their answer, they develop their understanding of the underlying math. At some point, when they represent the quantity 5, they will use tally marks such as ||||, or even begin to outline expressions such as ||| + ||.

ESSENTIAL

How students experience math is central to the Standards for Mathematical Practices, and trying to give students a deeper understanding of math is a central goal of the Common Core Standards for Mathematics.

Comparing Numbers

Pre-K students move from counting one group of items to counting two or more groups; they then compare the two counts. Being able to determine when two groups do and do not have equal amounts is a building block for answering questions such as, "How many need to be added to this pile to make them the same?" or "If I take two out of this pile, would the piles have the same amount?" Children can manipulate objects to figure out questions like, "How many would I need to take from the bigger pile, and put into the smaller pile for the piles to be the same?"

ALERT

A standard deck of playing cards *can* be great to use for comparing numbers, but some decks will be better than others. Examine the cards, such as the 8 of spades, to make sure that the symbols representing 8 spades is in a well-defined area, separated from symbols that may be near the number 8 to identify the suit as being spades. In many decks, counting the spades on an 8 of spades appears to have 10 spades on it.

Vocabulary for Comparing Quantities

Pre-K children build up their vocabulary by hearing and answering questions. You may notice a preference in how your child will answer questions. Encourage your child to vary his language. For example, when he is comparing the quantities of an item in two piles, he may have a tendency to always point to the larger pile and say, "That one has more." Encourage him to also observe (and communicate) that the other pile has less. Vocabulary used to identify smaller quantities includes *less than* and *fewer*. To identify larger quantities, children will learn to use phrases such as *greater*, *greater than*, and *more than*.

For groups with the identical amounts of items, children will use vocabulary such as *equal*, *equal to*, *same*, or *same as*. When the quantities are not the same, they will use phrases such as *different*, or *not equal*. Pre-K children begin to compare quantities by looking at groups of items, and comparing the size of the groups. Only after developing some counting skills and being able to represent a quantity with a number can students put meaning in comparing two numbers. How items are organized will affect how easily your child can count them. Kids learn first to count organized items, such as in a row. Then they can count items that are organized in rows *and* columns (or repeated rows). Finally, they can count items that are scattered. When children at this age are comparing groups of items, they may reorganize them to make them easier to count.

EXAMPLE 3

Create two unequal groups of pennies. Ask your child to compare the number of pennies in each group.

Discussion

As your child works on this exercise, he is both learning about relationships and practicing important vocabulary. Your child may use terms like *greater than*, *less than*, and *the same as*, or he may substitute other similar words.

Pose some "what if" questions to get your child to think about what happens when things change. "What if I take away a penny from this pile? What if I add a penny to this pile? How many pennies would I need to add (or subtract) from this pile so that they would have the same amount?" These are counting problems that can be acted out by physically adding and subtracting. It will help establish a connection between addition and having more of something, and between subtraction and having less of something.

If you're using coins that have two different sides, then you can teach about coins having heads and tails. Lay out a series of coins in the patterns shown here. Ask your child what comes next in each of the following patterns.

H T H T H

H T H H T T H H H T T

Extending patterns is an important skill. Have your child create a pattern and see if you can complete it. If your child has difficulty in seeing the difference between heads and tails, try this exercise with other objects, like blocks or balls.

Ordering, First and Last

Children enjoy organizing, sorting, and putting objects in order. At some point they will be able to take an unorganized list of numbers and put them in order. Pre-K skills focus on counting and being able to identify the *first* object and the *last* object in a list. Pre-K children focus on 1, 2, 3; in later grades students will identify items in an order as first, second, third. By focusing on first and last in pre-K, students will have a reference point for learning the numbers between.

EXAMPLE 4
Ask your child to draw a square around the first button in the row.

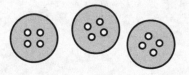

Discussion
The number of buttons may exceed the range that your child can count, but the exercise is to identify the first item in the row, regardless of how many items are in that row. What if she puts a square around the button on the far right. Is that wrong? Depending on how the row is oriented, perhaps

not; it's all a matter of perspective. However, because she should get used to reading from left to right, the best answer is the button on the left. Choosing a square as the shape to draw around the first object adds a little geometry to the question.

EXAMPLE 5

Ask your child to draw a circle around the last number cube in the row.

Discussion

Reading from left to right makes the number cube on the far right the *last* one. Choosing a circle as the shape to draw around the last object adds a little geometry to the question. This is another example of using subtle ways to work in extra vocabulary practice.

EXAMPLE 6

Ask your child if there are more buttons in the top row or in the bottom row.

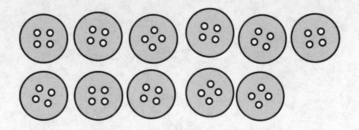

Discussion

Some children may just *see* that there are more in the top row. The follow-up question is "How do you know?" The objective is to count the number of buttons in each row, the top row totaling 6 and the bottom row totaling 5. The next step is to compare the numbers. Six is more than 5, the top row has 6, so the top row has more buttons. A visual method is to cross off one button in each row, and repeat until one row doesn't have any more buttons to cross out. Using this method, children can compare quantities beyond the range they can count.

Assessing Counting Skills

Use this checklist to assess the counting skills of your prekindergartener. Your child should be able to:

☐ Verbally count to 10

☐ Match a number of objects with a written numeral 0–5 (with 0 representing a count of no objects)

☐ Match numbers 1–10 with a corresponding set of objects or pictures

☐ When counting a group of items, identify the last number name said as the number of objects in the group

☐ State the number that comes before or after a specified number

☐ Identify the number in between when given a range of numbers, such as 3 and 5

☐ Count to answer "how many?" questions about as many as ten things arranged in a line, in a rectangular array, or a circle

☐ Count out objects for a given number from 1–10

☐ Compare two groups of objects using the terms *greater than*, *less than*, or *equal to*

☐ Identify *first* and *last* as related to order or position

☐ Duplicate and extend simple patterns using concrete objects

Addition and Subtraction

When your child becomes proficient with counting, he will realize that the organization of a collection of objects does not change the count. If five buttons are lined up in a row, the count is 5. When the buttons are scattered, the count remains 5. Then he learns that when additional buttons are added to or taken away from the group, the count will also change, regardless of how the objects are organized. As groups are changed and your child is asked for the new count, he may begin guessing or estimating, which are great problem-solving strategies even if they are not core skills. He can then count the items and see if his guess was correct.

Vocabulary for Adding and Subtracting

When presenting addition and subtraction problems to your child during routine daily activities, you provide clues as to the operation being performed using phrases that imply *more* or *less*. People *arrive*, increasing the number in the group (addition). People *leave*, meaning a number will decrease (subtraction). Children will use phrases like *totals*, *left over*, or *has*. The variety of vocabulary will help them to interpret real-world problems and improve their ability to break situations down into components they can understand.

Addition

Pre-K children develop their understanding of addition by using objects that they interact with. Problems that your child will perform on paper should be represented by pictures, not numbers. The idea is to get a visual understanding of a collection of objects, and understand that adding more objects will increase the total number of objects. The use of numbers should be restricted to counting the number of items in a group, associating the number with a concept instead of using the numbers in isolation. Children can memorize that $2 + 3 = 5$, but by memorizing these types of answers children lose the meaning. It is better for your child to see a group of objects, count them (2), see that more objects are being added, count those (3), and count all the objects after they have been combined (5).

ALERT

When creating addition stories, make sure that the sum remains within your child's countable range. If your child can count up to five, then make sure that the addition stories will not produce a number greater than five.

EXAMPLE 7

Sandra has 2 buttons lined up in a row.

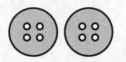

Barbara gives her 3 more.

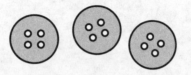

How many buttons does Sandra have?

Discussion

This example models the addition process: having a known amount, adding additional items, and getting a new total. This problem models $2 + 3 = 5$. The items are identical, they are organized, and they have quantities that are manageable. Here the total is five. As your child becomes proficient with adding numbers up through five, the totals can go higher. Early on, children are not expected to read these problems, but as they progress through the pre-K years they will be able to recognize the numbers as well as the words (2, 3, 5 being two, three, and five). In some resources you may encounter the word *sum* to describe the answer to an addition problem, but using *total* or *in all* are good word choices for building your child's vocabulary.

Subtraction

Pre-K children develop their understanding of subtraction by using objects that they interact with. Your child needs a visual understanding of having a collection of objects and know that taking some away will decrease the number of objects. Begin with a set of objects, or *manipulatives*, that she can count. Remove some from the group, and ask her to count the items that remain. Starting with a group that contains a count appropriate for your child will ensure that the answer will also be within her countable range. Problems that your child performs on paper could involve starting with a picture of a collection of items, with your child then asked to demonstrate "taking away" by crossing off some of them, and counting the items that have not been crossed off.

EXAMPLE 8

Sammy has 5 buttons lined up in a row.

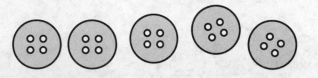

Sammy then gives Joey 2 buttons.

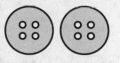

How many buttons does Sammy have left?

Discussion

This example models subtraction: having a known amount, removing a fixed amount, resulting in a *different* count. This problem models $5 - 3 = 2$. The items are identical, they are organized, and they have quantities that are manageable. Here the starting count is five. As your child becomes proficient with subtracting numbers beyond five, the starting counts can be increased. In looking through other resources you may encounter the word *difference* as being the term for the answer to a subtraction problem, but using *total*, *in all*, or *are left* are good choices for building your child's vocabulary.

Assessing Addition and Subtraction Skills

Use this checklist to assess the addition and subtraction skills of your prekindergartener. Your child should be able to:

- ☐ Create a verbal addition or subtraction story or scenario
- ☐ Describe and understand addition in terms of "all together"
- ☐ Describe and understand subtraction in terms of "taking away"

Measurement

Building vocabulary and being able to describe objects using age-appropriate words is what measurement is all about in pre-K. At the pre-K level, the units are not important. Students aren't learning to measure the weight of a rock, but rather comparing which of two rocks is heavier, and understanding weight as an attribute of an object.

There are of course times when your child might like to measure things and she will want to know units. For example, she may want to know how much she weighs, and she will use pounds as the unit. Or your child could want to see how tall she is and might want to know units, but she may prefer to use relative measurements such as "who's taller" or comparing markings on a wall to see "am I taller than I used to be?" Units of time or temperature may also be important to your child at this age. You can show your child that there are 5 minutes between each of the big numbers on an analog clock. If you are leaving in 5 minutes, she will be able to look at a clock and know when 5 minutes have elapsed and it is time to leave. This gives kids a positive feeling about knowing how to tell time, without them having to know how to actually tell time.

Vocabulary for Measuring Objects

When an object is measured, we use words to describe its length, width, or height. We use measurements to determine how long, short, or tall something is, and we can assign units to these measurements. Area can be described using terms such as *big*, *small*, *wide*, *narrow*, *fat*, *skinny*, *thick*, or *thin*.

Apart from describing an object's size and shape, we can look at other attributes such as color, shading, and if it is full or empty.

Vocabulary for Comparing Objects

When you and your child compare the attributes of two objects, you may use words such as *same* and *different*. Try introducing words like *longer*, *shorter*, *skinnier*, *fatter*, *thinner*, *lighter*, *heavier*, *bigger*, *smaller*, *wider*, and *narrower*. When you compare the attribute of three or more objects, vocabulary words such as *longest*, *shortest*, *skinniest*, *fattest*, *thinnest*, *lightest*, *heaviest*, *biggest*, *smallest*, *widest*, and *narrowest* are used. Using these vocabulary words regularly will help build your child's math, reading, and communication skills throughout elementary school and beyond.

EXAMPLE 9

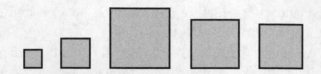

A. Which square is the biggest?

B. Which square is the smallest?

Discussion

It should be interesting to note how your child refers to the biggest square. The term "middle one" would be accurate of course, as he can see that there are two squares to the left and two squares to the right. He may count the squares and call the middle square number 3, which would be expected, or he may point to it. Hopefully he would identify the smallest square as the *first* one, but "square 1" is a good answer, too.

EXAMPLE 10

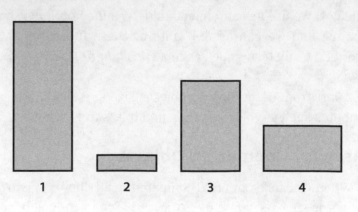

A. Which rectangle is the tallest?

B. Which rectangle is the longest?

Discussion

Questions about rectangles can be tricky. What are the attributes of a rectangle? Generally they will have a width and a length, but which is which to a child? The first question asked about being the tallest. So, if these are standing up, then rectangle 1 looks the tallest. If the second question asked for widest, would you have said rectangle 4? Which is the longest? If rectangle 1 were lying down, wouldn't it be longer than rectangle 4? Rectangle 3 and rectangle 4 are actually the same size, only rectangle 4 is lying down on its side. To determine which is longer, would it be fair to have all the rectangles oriented in their longest possible position for comparison? This illustrates some of the "fuzziness" about attributes and the words that kids will use to describe objects.

Assessing Measurement Skills

Use this checklist to assess the measurement skills of your prekindergartener. Your child should be able to:

- ☐ Identify measurable attributes of objects, such as length, and weight
- ☐ Accurately use vocabulary (e.g., *small*, *big*, *short*, *tall*, *empty*, *full*, *heavy*, and *light*) to describe the measurable attributes of objects
- ☐ Sort objects into categories
- ☐ Count the number of objects in each category (up to 10 objects)

Geometry

Children sort and order and classify shapes and objects based on common attributes as a normal act of play. They can be encouraged to look for specific commonalities or differences among both two-dimensional and three-dimensional shapes. They put shapes together to make other shapes (like two triangles of equal size creating a rectangle or square) and make constructions, such as house shapes. As they construct, they identify the formal spatial relationships between objects, such as *on top of* or *next to*.

ESSENTIAL

Young children often misidentify even the most common shapes, or fail to recognize similarly shaped objects, if the objects are not the same size. Unlike squares, whose shape remains the same as their size may vary, the shapes of triangles can vary greatly. Your child will come to understand that the attribute of "having three sides" determines whether a shape is a triangle, not how their side-lengths compare, as with a square.

Basic Shapes

Children will recognize the basic two-dimensional shapes: squares, circles, rectangles, and triangles, and may classify other shapes as these basic ones, even if they don't exactly match the attributes of these basic shapes. They may consider something that is oval to be a circle, which is reasonable as that may be the best match; they may consider a pentagon (five-sided figure) to be a square if the side lengths are all the same, or to be a rectangle if the shape looks more like a pencil.

Children may refer to any round three-dimensional shapes as a ball, and may identify any other three-dimensional shape as a box, or a cube. An age-appropriate name for a cylinder is a tube. They realize that the difference between two-dimensional and three-dimensional objects is that three-dimensional shapes can be filled up (they have volume). They develop a sense of what is a square and what may be a cube based on whether they can put stuff in it; they will decide whether a brick is shaped like a rectangle or whether a rectangle is shaped like a brick.

EXAMPLE 11

Draw a line from the shape on the top to the shape that looks the same on the bottom row.

Discussion

Part of sorting items is looking at how items are the same and how they are different. The intent of the question is to have your child examine the top shape, identify it as a triangle, and match it to the triangle on the bottom row. Your child could match the shapes based on how the interiors are shaded; children often find their own ways to match and sort items. Working with a child on this question uses the vocabulary words *line*, *top*, and *bottom*. Problems like this can also be extended to ask a child to identify the first or last shape in the bottom row; you could also ask him to draw a circle or square around the first or last item in the row. These extension questions help to develop his spatial awareness.

Complex Shapes and Constructions

Pre-K students begin to use and understand vocabulary that represents spatial relationships while constructing two-dimensional shapes or three-dimensional constructions, or by putting objects in order. Young children may classify or group objects based on one particular attribute, such as color or shape. As students get older they may sort objects by multiple attributes, such as first dividing a collection by shape, and then by color. These early-learning skills are used throughout their elementary education, as they learn an increasing number of attributes.

Vocabulary Describing Position

While completing a puzzle, your child may use the phrase *goes there* to mean that a piece goes between two other pieces, or goes under one piece and on top of another. When he plays with blocks he will put one in front of or behind another. Other prepositions he may use include *beside*, *inside*, *outside*, or use directions such as *left*, *right*, or *middle*. Pre-K students will use spatial distance terms such as *near*, *far*, *close to*, *far away*, and *farther*.

EXAMPLE 12

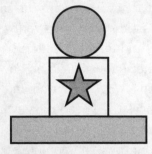

A. What shape is in the middle of the square?

B. What shape is under the square?

C. What shape is on top of the square?

Discussion

Children answering these types of questions think about the names of the shapes and their attributes, and the spatial relationships between the shapes. The question here centers around the square, and asks about shapes that are *on top*, *under*, and *in the middle*. Your child will have to recognize the rectangle and the circle as basic shapes, and may or may not know the name for the star.

Assessing Geometry Skills

Use this checklist to assess the geometry skills of your prekindergartener. Your child should be able to:

- ☐ Use the names of shapes to describe objects around them
- ☐ Describe the relative positions of objects using terms such as *top, bottom, up, down, in front of, behind, over, under*, and *next to*
- ☐ Correctly name shapes, regardless of size
- ☐ Compare and sort two- and three-dimensional shapes and objects, in different sizes, based on their similarities
- ☐ Compare and sort two- and three-dimensional shapes and objects, in different sizes, based on their differences
- ☐ Create and build shapes from components (e.g., sticks and clay balls, blocks)

Helping Your Child Succeed

How can you help your prekindergartener with her math skills? Integrate counting things into your daily routine. Use manipulatives to model counting: pennies, buttons, small blocks, polished rocks, stickers, or maybe small crackers. Vary the objects to make the activities interesting to your child. Counting can be a fun way to introduce math to young children, and it builds their self-esteem as they learn to count higher and higher. Use the checklists provided at the end of the previous sections to monitor the growth of your pre-K student's math skills.

Also, read out loud with your child. Communication is a crucial skill; math problem solving will be more word-intensive than it has been in the past, and solving real-world (word) problems are now valued more than calculation skills. Students in later grades will need proficient reading and writing skills; the student who struggles with reading and reading comprehension can find that math becomes less accessible. Children need to make the connection between the words that they hear and words that they read.

HOME ACTIVITY 1

Count up to five using your fingers; start with a closed fist and show one finger at a time, counting out loud. After mastering counting to five, start with five fingers shown and take down one finger at a time, until you have zero fingers shown and your hand returns to being a closed fist.

Discussion

Here your child makes a connection between the word representing an amount and the number of fingers shown. Indirectly, you are demonstrating "adding one." Zero is an important concept, but it isn't part of the normal counting process. If you are counting items, it is unlikely that you would start with zero, that's why the standards specify counting from 1. In the classroom, when children are asked to count the number of objects in an area, they will not be shown 0 items. Pre-K teachers teach about zero by counting down how many things remain after items are taken from a group, until there are zero items left. This becomes an important concept when students are learning subtraction.

After mastering counting forward to and backward from five, try these activities:

- Pick a number from 1 to 5 at random, and have your child hold up that number of fingers
- Count up to 10 fingers, and then down starting from 10 fingers
- Hold up a random number of fingers and ask how many fingers are being shown
- Hold up a random number of fingers and begin counting up to five (or ten) from that point
- Hold up a random number of fingers and begin counting down from that point

HOME ACTIVITY 2

Create addition and subtraction stories. Tell your child that there is some quantity of objects, and some quantity is added or removed. Then ask your child to tell you what the new quantity would be. For example:

{Person's name} has {some number} of {item(s)}. {Another person's name} brings {another number} more. How many {items} are there now?

These could be verbal problems, and could even become part of your end-of-day routine, or you could work together to create storybooks with drawings or pictures cut from magazines.

Discussion

Children are very creative; you could prompt your child to provide you with a number to start with and an object to be counted. You could have him supply you with names for people who bring things to add to the group, or to give things to from the group. You could have him create a story for you to solve and could have him check your answer.

HOME ACTIVITY 3

Use a set of index cards (or cut pieces of paper) to create a set of sort-match cards. A sort-match activity takes a deck of cards and asks students to organize them into related sets. For the following sample cards, from the Counting to 10 deck, students are asked to match the number of stars on each card with the numeric representation and the verbal (word) representation. For students not yet ready for the verbal model, all of the text cards can be removed from the deck.

1	one	★
2	two	★ ★
3	three	★ ★ ★

Begin with 15 index cards. Take 5 cards and place a large numeral 1 through 5 on each card; take another group of 5 cards and write the word for each of the numbers, one through five, on each; then choose an object that you can draw (or use stickers) and create cards with 1 through 5 objects on each. This can be repeated for the numbers 6–10 as desired.

Mix the cards up and have your child create sets of three cards. Each set includes one number card, one word card, and one picture card; each card in the set must represent the same quantity. The game ends when there are five correct sets.

Discussion

The Standards for Mathematical Practice emphasize the importance of knowing multiple representations for math concepts. Here your child uses quantities represented numerically, pictorially, and verbally. This activity will provide you with an observational assessment; can your child count to five? Or ten? Children will also begin to understand patterns as they explore this set activity; is it possible to have only one incorrect set of cards?

HOME ACTIVITY 4

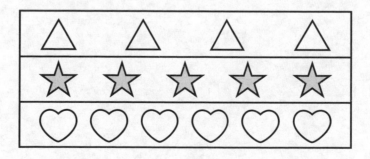

Tell your child to draw a circle around the first object in each row, and draw a square around the last object in each row.

HOME ACTIVITY 5

Use buttons or any other small items you have on hand to illustrate this addition story:

> Leigh has 6 buttons lined up in a row. Doug then gives Leigh 3 more buttons. How many buttons does Leigh have?

HOME ACTIVITY 6

Use small items to illustrate this subtraction story:

> Lisa is playing a game with 5 number cubes. Lisa then lets Lou play with 2 of them. How many does Lisa have left?

CHAPTER 4

Kindergarten

Kindergarten is an exciting time, full of new experiences and myriad new things to learn. In kindergarten math your child will use sets of objects for counting, comparing, and modeling addition and subtraction. She will learn to represent quantities with numbers up to 20, and how to count up to 100. In kindergarten she'll also describe basic two-dimensional and three-dimensional shapes in the world around her. She'll describe the special relationships between objects, such as *next to* and *on top of*, and use basic shapes to construct more complex shapes.

What Your Kindergartener Is Expected to Learn

In kindergarten your child will count, and count, and count! Kindergarteners count objects in a group, and learn to represent the quantity of items with a number. This connection between a number and a quantity of objects is an important basic math skill used throughout elementary school. Students in kindergarten will also:

- Learn that addition is "having some quantity of objects and getting more," and that "adding will result in having more in total."
- Physically model addition and subtraction problems using objects they can count, and they will learn multiple strategies for counting and performing addition and subtraction operations.
- Observe the *commutative property* using sets and numbers such as $3+4=4+3$.
- Find different ways to break up quantities less than 10, such as seeing that 7 is the same as both $5+2$ and $6+1$.
- Learn to see numbers 11 to 19 as a group of ten and some leftover ones, which provides a foundation for learning place value in future grades.

This chapter will help you understand the types of problems your child encounters in kindergarten, and the learning outcomes that the Common Core Standards identify.

FACT

The *commutative property of operations* says that the order of the numbers does not matter, so $7+8=15$ and $8+7=15$. In other words, the numbers can be rearranged and not affect the sum.

Common Core State Standards: Kindergarten

There are five general categories of math skills your child will be learning in kindergarten, and certain abilities that fall under each of those categories:

COUNTING AND CARDINALITY
- Know number names and the count sequence
- Count to tell the number of objects
- Compare numbers

OPERATIONS AND ALGEBRAIC THINKING
- Understand addition as putting together and adding to, and understand subtraction as taking apart and taking from

NUMBER AND OPERATIONS IN BASE TEN
- Work with numbers 11–19 to gain foundations for place value

MEASUREMENT AND DATA
- Describe and compare measurable attributes
- Classify objects and count the number of objects in categories

GEOMETRY
- Identify and describe shapes
- Analyze, compare, create, and compose shapes

For the full list of Common Core State Standards for Mathematics, see Appendix A.

What Your Child May Know Before Kindergarten

Preschool and formal prekindergarten programs and opportunities vary from state to state, so rather than specifying skills that incoming kindergarteners *must* know, here is a list of skills they *may* already have. Your child may be able to:

- ☐ Verbally count to 10
- ☐ Match a number of objects with a written numeral 0–5 (with 0 representing a count of no objects)
- ☐ Identify "first" and "last" as related to order or position
- ☐ Use the names of shapes (circle, square, triangle, and rectangle) to describe objects around them

Counting and Cardinality

Kindergarten counting begins by ensuring that students are able to successfully count to 10, and write the numbers to 10 and then to 20. Students are learning *how* to count at the same time they are learning the number names. Many students may have been previously exposed to counting to 20 and are ready to count well beyond. Students at this age will often be able to count to higher numbers than they are able to write.

In kindergarten your child will be expected to count up to 20 objects within a set, or collection of objects. While counting, your child will use a one-to-one correspondence from a number to a specific object in the set. For example, he may point to each item as he counts and realize that the number of items in the set is the same as the number he said aloud for the final item counted. He will be able to use counting skills to answer "how many?" questions. Your child will use several strategies to compare the number of items in two sets, and use expressions such as *greater than*, or *less than*, or *equal*. He will be able to sort using numbers, or by organizing collections of objects from sets that have fewer objects to sets that have more objects.

FACT

Cardinality is a measure of how many things are in a group. For example, if group A has a star, a circle, and a square in it, then the cardinality of group A is 3. Your kindergartener will not actually use the term *cardinality*, just learn the concept it expresses.

Number Names and Sequence

At the end of kindergarten students will be able to count to 100 by ones and by tens. When students really understand the sequence of numbers from 1 to 100, they will be able to start at any number and count up to 100. It is common for students learning to count to always start at 1, even when asked to begin a sequence from a higher number. A good progression for students is to first count to 10, then to 20, and then be able to start from any number and count up to 20. Zero is not usually part of a counting sequence, and your child will not often be asked to count a quantity of zero objects. Zero is used for counting down, and with subtracting items until there are 0 left.

EXAMPLE 1

Fill in the missing numbers, 1 to 10

1   2   ___   4   5   6   ___   8   9   ___

Fill in the missing numbers, 1 to 20

1   2   3   4   5   6   7   8   9   10

11   ___   ___   14   ___   16   17   ___   19   20

Fill in the missing numbers, 1 to 100

1   2   3   4   5   ___   7   8   9   10

___   12   13   ___   15   16   17   18   19   20

21   22   23   24   25   ___   27   28   29   30

31   32   ___   34   35   36   37   ___   39   40

41   ___   43   44   ___   46   47   ___   ___   50

___   52   53   54   ___   56   57   58   59   60

61   62   ___   64   65   66   67   ___   69   70

71   72   73   74   75   76   77   ___   ___   ___

___   ___   ___   84   85   ___   87   88   89   90

91   92   93   ___   ___   ___   ___   ___   99   100

Discussion

When students see counting charts like these, they find patterns that help them understand the relationships between the previous and following numbers in a row, and see patterns throughout the columns. The tables can be varied by omitting patterns of numbers; for example, removing the 10s to help your child count by tens.

Another good way to improve number sequence skills is playing the I Say, You Say game. In this game, you can ask your child questions like these:

- I say 11, you say the number that is one more.
- I say 44, you say the number that comes next.
- I say 77, you say the number that comes next.
- I say 25, you say the number that is one more.

By varying the way questions are asked, such as "the number that is one more" and "the number that comes next," your child will understand that each number in the counting sequence is a quantity that represents "one more." (This addresses a specific kindergarten standard.) You can also vary the game to I Say, You Write with similar questions for numbers 1 to 20, but have your child write the answers.

Counting Objects

When items are neatly in a row, like the following buttons, they are easier to count.

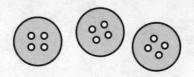

When items are in a more scattered formation they can be more difficult to count. The Common Core Standards call for students to be able to count up to 20 items that are well organized, such as in a line, or a circle, or a rectangular array.

ESSENTIAL

Children in kindergarten will use the term *array*, repeating the vocabulary they hear at school. The term *rectangular array* refers to items being organized in rows and columns.

However, the standards also call for students to answer "how many?" questions for up to 10 items in a *scattered* configuration. Students using physical objects will probably be able to count more than 10 items in a scattered configuration by first organizing them into more manageable (and countable) configurations, but this solution is not always possible for paper or electronic activities where 10 items is a more practical limit.

EXAMPLE 2

Count the number of buttons.

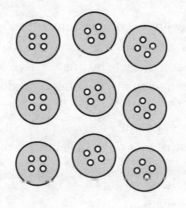

EXAMPLE 3

Count the number of buttons.

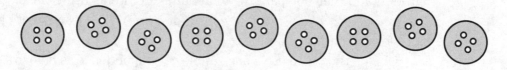

EXAMPLE 4

Count the number of buttons.

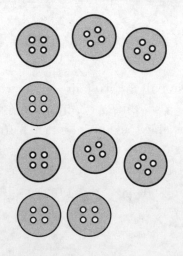

Discussion

Experienced readers may count each row from left to right, to count 9 buttons in each collection. Adult readers may view the 3×3 array as more of a multiplication problem and "see" that the answer is 9. Kindergarteners may count any of these in what appears to be a more random order, maybe left to right, then right to left, or up and down. The important point is to make sure that your child uses one number to count one item when counting, with some structure to ensure that no item is missed or counted twice, and that the count is accurate.

Comparing Numbers

After your child is able to count the number of items in a group, the next skill he will learn is to be able to count the number of items in two groups, and then compare the total number of items using vocabulary such as *greater than, less than,* or *equal* (or *same*). He'll learn multiple strategies for comparing quantities.

One method is to use a one-to-one correspondence as he pairs one item from one group to one item in another group. If he "runs out" of items in one group before the other, he can identify which group has fewer, and which group has more (or a greater number). If every item in one group is paired with an item in another group, and there are no leftover items in either group, then the quantities are equal. An advantage of this method is that your child can compare higher quantities than he is able to count. Students are expected to be able to compare two groups of up to 10 items per group.

Another method of comparing groups of items is to count the items in one group, then count the number of items in the other group, and then compare the numbers. Students less proficient with comparing two numbers can plot both numbers on a number line to visually see which number is less than, or greater than the other number.

EXAMPLE 5

Angela has a group of beads. Clara has a group of erasers.
Who has the most?

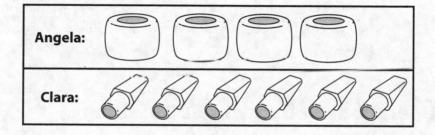

Discussion

The question presents two collections and does not suggest a method for comparing the two groups. The way the collections are organized, one above and one below, may invite students to cross off pairs, or circle pairs to see who has the most.

EXAMPLE 6

Circle the greater number.

A. 4   7

B. 5   3

C. 8   2

Discussion

It is often easier for kindergarteners to identify the lesser of two numbers than the greater of two numbers. Students often begin counting from 1 and realize that the first number they come to in the counting sequence is the lower number, so the greater number must be the other one. Alternatively, if they are looking for the greater number they could start from 10 and count backward, and the first number they encounter is the greater number. When students become proficient with comparing two numbers from 0 to 10 they will be able to verbally answer comparison questions, and be ready to include numbers greater than 10.

ESSENTIAL

Students are not expected to provide detailed explanations, but they express their understanding using precise vocabulary phrases such as *less than*, *greater than*, or *equal* (which addresses the Mathematical Practice of Attending to Precision). The question types that students see will help them express their answers in a variety of formats, such as through vocabulary, pictures, or numbers.

Operations and Algebraic Thinking

Kindergarteners perform addition and subtraction. They are engaged in math by having a collection of objects and either adding to it or taking things away from it. They learn to connect addition to the operation of getting more, and subtraction as the operation that results with having less. By using hands-on activities to move objects in and out of groups and by combining groups, your child builds up arithmetic fluency. In kindergarten your child focuses on addition and subtraction up to the number 5, and later will use addition and subtraction problems up to 10 to prepare for first grade.

EXAMPLE 7

Add.

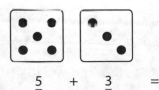

<u>5</u> + <u>3</u> =

Discussion

In kindergarten your child will count sets of objects and find a total. She may do this by drawing the answer, such as 8 dots, or her teacher may lead her through the problem to use numerals. A problem such as this can be used as a steppingstone from performing addition using strictly pictures to performing addition using strictly numbers. Your child may see this as an *and* operation, explaining that 5 *and* 3 is the same amount as 8.

EXAMPLE 8

Chris put 3 blue pens and 3 red pens in a bag. How many pens did Chris put in the bag?

Discussion

Kindergarten word problems use age-appropriate vocabulary. You will notice problems such as this one involve two colors of the same type of object. The question focuses on the number of pens in the bag, and not the colors. Even at early ages children need to identify the attributes that are important to a question. Parents of children struggling with vocabulary and math can use words familiar to their children to ease anxiety of word problems. Addition problems are based on having some quantity of items and adding more items, or putting together two quantities of items. The focus is not on converting a word problem to a number sentence (such as $3+3=6$) and then solving the number sentence, it is understanding the operation being performed (addition or subtraction).

EXAMPLE 9

Carla put 4 apples in a bag. How many more apples does Carla need to have 10 apples in the bag?

Discussion

At some point your child may recognize this as a subtraction problem, but in kindergarten this combines the skills of being able to count to 10 starting from any number less than 10, and counting how many items are added to the starting quantity—in this case, 4 apples. You can help your child with problems like these by using small objects (known in the math world as *manipulatives*) to illustrate the concept. Start with a group of 4 and tell your child to add a new group, one item at a time, until she reaches 10. Then she can count the number of items added to the new group to answer the question. She would be modeling the problem $4 + ? = 10$.

ESSENTIAL

EXAMPLE 10

Carla has 10 apples. She gives away 4 apples. How many apples are left?

Discussion

Your child can act out this problem by starting with 10 apples (or counters), taking away one at a time until 4 have been removed, and then counting the remaining apples. She is learning that one quantity is being reduced by another quantity, and that this reduction process is what subtraction means.

In kindergarten students are learning about the space around objects and use vocabulary like *behind* and *next to*. They can be confused about the use of the word *left* in a question such as this, especially when objects used in addition and subtraction problems are neatly organized. As an adult, you understand that the word *left* in this context is used to mean *remain*, but to a child this can lead to confusion. Having your child explain her answer will help you to assess her understanding far more than just examining her answer.

FACT

Picture-based subtraction problems on paper can't be moved around to demonstrate an operation like objects can. But you can show your child how to cross off or mark items that are being taken away. For example:

EXAMPLE 11

Owen saw 6 fish in a pond then 2 fish swam away. How many are left?

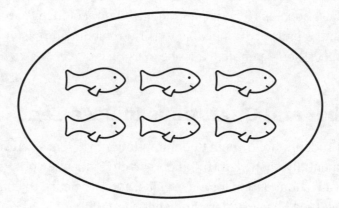

Discussion

Again, make sure your child understands that *left* has multiple meanings: Two fish left, but there are four fish left. To solve this problem, your child might mark the two fish that swam away, and then count the fish that did not swim away:

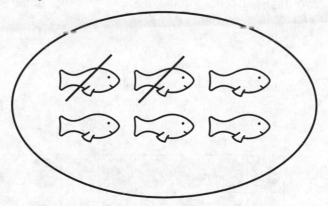

If your child is proficient using number pairs, such as the number pairs that add up to 10, he might see how it can apply to solving subtraction problems.

EXAMPLE 12

Katie had 10 pennies, but now she has only 6. How many pennies did she lose?

Discussion

Part of the Operations and Algebraic Thinking standards promote being able to *decompose* (break down) numbers (such as 10) into pairs of numbers such as (6 and 4). Students who recognize that the 6 and 4 pair (or 4 and 6 pair) are applicable to this problem will reason that if she has 6 pennies, then 4 pennies must be missing.

Number and Operations in Base Ten

Students build their understanding of the base ten number system incrementally each year. In kindergarten, students work with numbers 11–19 to gain a foundation for place value, and they learn to count to 100 by ones and by tens. They understand that the numbers 11 through 19 represent one group of ten plus some additional ones.

The *base ten number system* refers to the position of numbers. For example, take the number 265. The 2 is in the hundreds position, the 6 is in the tens position, and the 5 is in the ones position. Each number-place is ten times the value to the right of it (hence the term "base ten").

Students in kindergarten (and first grade) may use filled and partially filled ten-frames, such as this one representing 19, to represent numbers, quantities, addition, and subtraction operations.

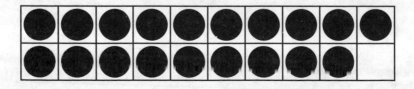

Students in kindergarten could be presented with two blank ten-frames and asked to represent any number from 11 to 19, or asked to identify a number (from 11 to 19) from partially filled ten-frames.

EXAMPLE 13

Show an amount the same as 15.

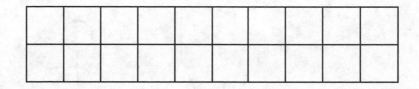

Discussion

The target skill for this example is having your child decompose 15 to $10 + 5$ by filling in the ten-frame on the top completely to represent 10, and filling in 5 of the available frames on the bottom. That may take some coaching, as your child may see that filling in 8 frames in one and 7 in the other is also 15, or she may just fill in 15 frames at random from the available twenty. While she would not be incorrect in doing so, the focus of this exercise is having your child see numbers 11–19 as completely filling in a ten-frame, and then having some leftovers representing "ones."

Measurement and Data

The attributes measured in kindergarten are usually *length*, *width*, *height*, and *weight*. Your child will work on answering comparison questions such as "which is longer" (or shorter, or lighter, or heavier). He experiences measurement by putting objects next to each other, and comparing similar attributes. He orients objects so that they can be compared fairly. At the kindergarten level, the emphasis is on the comparison, not on measuring attributes in exact units such as inches, feet, or pounds. Your child will classify objects and count the number of objects in each category. For example, this is the type of measurement problem your child might see:

Who is taller? Complete the sentence: _____ is shorter than _____.

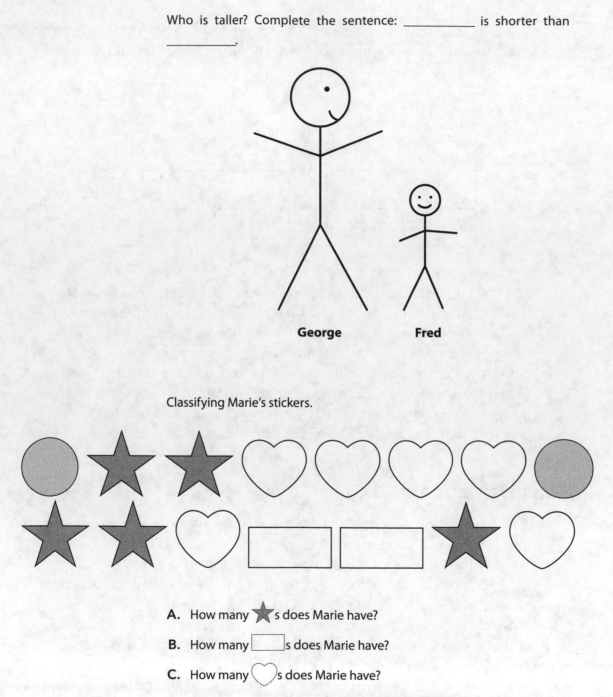

George   **Fred**

Classifying Marie's stickers.

A. How many ⭐s does Marie have?

B. How many ▭s does Marie have?

C. How many ♡s does Marie have?

D. Put these numbers in order from least to greatest.

Measurable Attributes

Measurable attributes can be assigned units, but in kindergarten often the units will be in terms of other objects. For example, two lengths of connected blocks could be measured using the number of blocks, but not necessarily measured in inches or feet. By comparing attributes, students will be able to describe how two objects are similar and how they are different. Measuring or comparing attributes helps with classifying and sorting objects.

EXAMPLE 14

Which tree is taller?

Discussion

What is the thought and reasoning process that a child would use to solve this problem? If he draws a straight line from the top of the tree on the right over to the tree on the left, he will notice that the line meets the tree on the left about halfway up. From this he may consider the tree on the right to be one unit tall, and that the tree on the left is taller than one unit. Similarly, if the tree on the left represents one unit, when compared to it the tree on the right is less than one unit tall. Questions that ask "Which is taller?" should be followed with "How do you know?" in order to assess understanding and also develop communication skills.

EXAMPLE 15

Which stack of blocks is bigger?

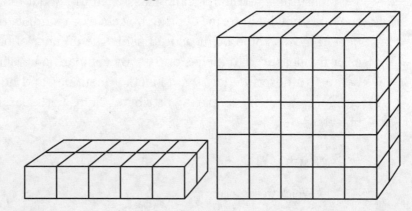

Discussion

Your child may answer this question right away, or may consider that many of the attributes of these two stacks are similar. They both have 5 blocks along the bottom, and they are both 2 rows thick; does *bigger* in this case mean the same as taller? *Bigger* could compare the surface area; treated as two piles of same-sized shapes, *bigger* could mean more, and the pile on the right is therefore bigger because there are more cubes in the pile.

Classifying Objects

An interesting activity at any grade level is to provide your child with a bag of random objects, without providing any directions. What would your child do? Probably take the objects out of the bag, and sort them into categories. Children determine for themselves the sort criterion used, which could be based on size, shape, weight, or other attributes that can be measured; sorting could also be done by color, or other attributes that are difficult to measure but not difficult to compare.

When your kindergartener is asked to classify objects, the number of objects in any one category should be less than or equal to 10. Electronic activities asking your child to classify objects can be engaging with stories, sounds, and animation. You can create engaging and interesting activities for your child to classify physical objects found around the house. Your child will also encounter paper-based activities asking her to classify objects in a variety of ways.

SAMPLE PAPER-BASED CLASSIFICATION EXERCISES

- Your child could be asked, using a multiple-choice format question, to identify which two shapes are alike from within a small collection of shapes; or she may be asked to identify the one that is different. The method for identification could be circling the shape, or marking a box, or selecting a number or letter that corresponds to the shape. These questions are important for building vocabulary as much as being able to classify objects.

- Many times the classifications that kindergarteners perform are based on separating objects by how they are used, or who would use them, or where they would be found. For example, a whistle and a badge are something you would expect police to carry, while a hammer and measuring tape are more appropriated for a carpenter. A measuring cup would be for a baker.

- Classification on two-dimensional objects can be based on the number of sides or colors.

- Classification of everyday three dimensional objects can be sorted by general cylinder, cube, cone, or spherical shape. Your child could also be asked to classify objects as being either 2-dimensional or 3-dimenstional.

- Once objects are classified, or grouped, each group is counted, and the "cardinality" of each set becomes an attribute of the set, so the sets of objects can be compared. Vocabulary used to prompt students include: which group has "the most," "the greatest number," "the fewest," the least," and "the lowest number."

Geometry

That's right, even in kindergarten your child delves into geometry! The basic two-dimensional shapes your child encounters in kindergarten include squares, rectangles, triangles, circles, and hexagons. The basic three-dimensional shapes your kindergartener learns about are cubes, cones, cylinders, and spheres. Kindergarteners will often use more comfortable names to describe the three-dimensional shapes, such as tubes or balls. Students explore working with three-dimensional shapes to make new three-dimensional shapes.

Basic Shapes

Children will look at the area around them and connect the objects they see with shapes they can identify. A book is a rectangle or a square, a sandwich may be cut into rectangles or triangles, and coins are circular. Students are asked to classify objects as being *flat* or being a *solid*, and could encounter the terms *two-dimensional* and *three-dimensional*. In kindergarten, students develop their sense of special relationships and are asked to describe the relative positions of objects. For example, are the shoes *under* the table, or *on top of* the table? They discuss position as a function of perspective—for example, a chair can be seen as being in front of a table from one view, and behind a table from another view.

EXAMPLE 16

Ask your child to identify the basic shapes indicated by the Common Core Kindergarten Standards.

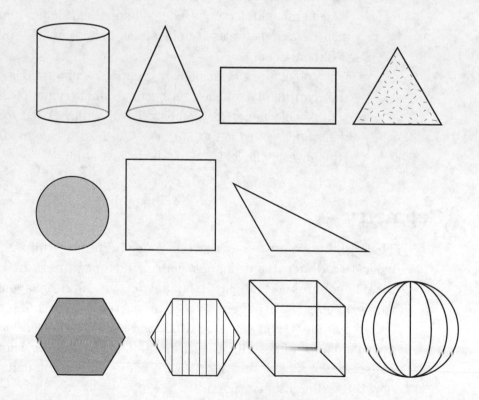

Discussion

Talk with your child about the attributes of shapes. Triangles have three sides; how the triangle is filled in, the size of the triangle, or how it is oriented (turned) does not change it from being a triangle. Triangles, unlike circles and squares, can change appearance: Their angle measurements change and their side lengths vary. Discuss with him that rectangles always have four sides and four corners. Also have him differentiate between flat shapes and three-dimensional solids.

Working with Shapes

Children can take two triangles and put them together to make a square. Maybe they use scissors and cut a square into two rectangles, or take two rectangles and create a T-shape. Students compose new shapes from basic shapes, and explore how using multiple same-sized shapes can (or maybe cannot) be used to create larger versions of the same shape. Can you use multiple rectangles to create a larger rectangle? Can you use multiple circles to create a larger circle? Can you use two basic shapes to create another basic shape?

As students use both two- and three-dimensional shapes they will explore, analyze, compare, create, compose, and construct shapes and develop understanding about similarities, differences and attributes. They will see that a triangle always has three sides, regardless of how big or small it is or whether it was created using four or fifteen smaller triangles.

EXAMPLE 17

Cut out four equal-sized squares from an index card or colored paper. On a separate sheet of paper, see how many different shapes can be made by placing the squares side by side, using a varying number of squares. There is only one shape that can be made using a single square, and one shape that can be made using two squares. Try using three squares, and then four squares. How many shapes can you make?

Discussion

Two shapes can be made using three squares, and five different shapes can be made using four squares. If you are ambitious, there are twelve shapes that can be made using five squares. Remember that the sides of the two squares must be touching, not just the corners. Discuss how the shape your child constructs doesn't change just because it is oriented differently.

ESSENTIAL

Blokus is a great educational family game where players take turns fitting two-dimensional shapes on a playing board. The shape pieces are constructed similarly to Example 17, using up to five same-sized squares. A 3D version is also available.

Helping Your Child Succeed

When you help your child learn to count, and to understand addition as an operation that increases a quantity and subtraction is an operation that decreases a quantity, you help her not only in kindergarten, but also through first grade and beyond. Continually assess your child's progress throughout her kindergarten year by asking her questions involving counting, addition, and subtraction each week. Encourage your child whenever she struggles, but allow time for her to think through some challenges. Praise your child for persevering through anything challenging.

Around the House

Show your child that opportunities to count can be found anywhere. Have her count sets of between one to ten items, and then eleven to twenty items. Have her count items out loud, saying the numbers as she counts each item, using a one-to-one correspondence between each number name and one (and only one) object. Then have her try to write the numbers.

Create a set of sort-match cards. Using index cards, make three different versions depicting numbers 0–20. One card in each set shows the numerals (1, 2, 3); one shows the corresponding numbers of objects (circles, stars, apples); and the last set shows the number words (one, two, three). You may select cards from the deck that are appropriate for the development level of your child, maybe cards 1–5 at first, then 0–10, and then 0–20. The sort-match exercise requires your child to represent each quantity in three ways. As a first step, use the cards in sequential sets. As your child progresses, use the cards in random order.

Here are some other sort-match activities you can try:

- Select cards at random and ask your child to identify the number (or the quantity). Then ask her to count to that number, or from that number up to 10 (or 20, or 100).
- Ask your child to sort the 1–5 (or 1–10 or 1–20) picture cards from lowest to highest (or highest to lowest). Once she's mastered this, try it with the numeral cards, then the word cards. (Keep in mind that using all of the cards in this way exceeds the Common Core Standards for Kindergarten.)
- Select pairs of cards to use as memory/concentration games.

Illustrative Mathematics (*www.illustrativemathematics.org*) has lots of counting and sequencing games you can play with your child.

Try these:

- Start counting at 1 and I will tell you when to stop. (Stop at 22.)
- Start counting at 10 and I will tell you when to stop. (Stop at 35.)
- Start counting at 54 and I will tell you when to stop. (Stop at 68.)
- Start counting at 86 and I will tell you when to stop. (Stop at 102.)
- Tell me the number after 2. After 5? After 8?
- Tell me the number after 10. After 13? After 16?
- Tell me the number after 20. After 24? After 29?
- Tell me the number after 55. After 79? After 87?

KindergartenKindergarten.com (*www.kindergartenkindergarten .com*) is a great site for wonderful activity ideas you can do around the house.

CHAPTER 5

Grade 1

First grade, the big time, the world of the big kids. Entering first grade is quite an achievement for your young one. She is most likely comfortable with the school setting, and is ready for new learning experiences. In first grade math, your child will develop multiple strategies for addition and subtraction up to 20, and develop an understanding of place value that will help her add numbers up to 100. She will develop her understanding of measuring lengths, and reason about attributes of shapes. This chapter presents the types of problems that your child will see in first grade and discusses some of the strategies she will learn including drawing diagrams to model a problem, using number lines, and completing equations.

What Your First Grader Is Expected to Learn

At the end of first grade your child will be able to do the following:

- Count, read, and write numbers through 120
- Add numbers to create tens, then use skills for grouping tens to help him add single- and two-digit numbers up to 100
- Add up to three two-digit numbers
- Compare numbers using the relational operators *less than*, *greater than*, and *equal*
- Develop an understanding of an equal sign as representing two equal quantities
- Determine whether an equation is true or false
- Model word problems, addition, and subtraction problems using diagrams, number lines, and equations
- Add and subtract numbers within 5, then within 10, and then within 20
- Using addition, for example, your child sees the similarities in the structure of $5+2$ and $2+5$ and apply the commutative property without having it formally introduced
- When he adds three numbers together he may regroup them, combining numbers to first find groups of tens to simplify calculations

In first grade your child will use graphs, such as bar graphs, to represent up to three categories of objects. He'll also learn to read graphs to answer questions, such as comparing the number of objects in each category.

In terms of measurement, your child will measure lengths using standard units such as inches or centimeters, but he may also measure and compare two objects in terms of a third object. He will be able to order objects from shortest to longest, or longest to shortest.

That is a lot to learn in a year! Let's look into each of the skills your child will learn and examine how you can help build those skills at home.

Common Core State Standards: Grade 1

There are four general categories of math skills your child will be learning in first grade and certain abilities that fall under each of those categories:

OPERATIONS AND ALGEBRAIC THINKING

- Represent and solve problems involving addition and subtraction.
- Understand and apply properties of operations and the relationship between addition and subtraction.
- Add and subtract within 20.
- Work with addition and subtraction equations.

NUMBER AND OPERATIONS IN BASE TEN

- Extend the counting sequence.
- Understand place value.
- Use place value understanding and properties of operations to add and subtract.

MEASUREMENT AND DATA

- Measure lengths indirectly and by iterating length units.
- Tell and write time.
- Represent and interpret data.

GEOMETRY

- Reason with shapes and their attributes.

For the full list of Common Core State Standards for Mathematics, see Appendix A.

What Your Child Should Know Before First Grade

In kindergarten, your child learned to count to 100, by ones and by tens. This will help him learn to count beyond 100 in first grade, and to use tens to further his understanding of place value. Your child should be able to represent a quantity of objects with a number, and be able to write and recognize the numbers from 0 to 20. He should be able to compare the size of two groups of objects to see if they are equal, or be able to determine which group is greater than or less than the other. Upon entering first grade, your child should already able to fluently add and subtract within 5, and will build upon those skills to add and subtract within 10. Also, he should be able to name basic two-dimensional (flat) shapes, and three-dimensional (solid) shapes.

Operations and Algebraic Thinking

In first grade, your child will be presented with word problems, as well as equations where some parts of the problem are given and one part is unknown.

Addition

Consider the following equation:

$$2 + 3 = 5$$

In this equation, 2 can be viewed as the starting number and 3 is the amount of the change. Five is the new amount, the sum of the starting number and the change. The Common Core Standards use this example for modeling three primary types of addition word problems:

- **The result is unknown:** Two bunnies sat on the grass. Three more bunnies hopped there. How many bunnies are on the grass now? ($2 + 3 = ?$)
- **The change is unknown:** Two bunnies were sitting on the grass. Some more bunnies hopped there. Then there were five bunnies. How many bunnies hopped over to the first two? ($2 + ? = 5$)
- **The start is unknown:** Some bunnies were sitting on the grass. Three more bunnies hopped there. Then there were five bunnies. How many bunnies were on the grass before? ($? + 3 = 5$)

First-grade students will answer all three types of these questions. It is expected that at the end of first grade, students will be proficient with answering questions where either the result is unknown, or the change is unknown. First-grade students are expected to answer addition word problems within 20, and be able to work with up to three addends.

FACT

Addends are simply the numbers being added. In $2 + 3 = 5$, the numerals 2 and 3 are the addends. In $a + b = c$, a and b are considered the addends.

EXAMPLE 1

Solve the word problem using a picture, a number line, or an equation.

Ricky earned 5 stars on Monday from his teacher, 4 stars on Tuesday, and 6 stars on Wednesday. How many stars did Ricky earn?

Discussion

On a number line, either 0 or the number 5 could be identified as the starting point, with additional jumps of 4 and 6 being represented for earning stars on subsequent days. 15 would be identified on the number line as the answer. Using a diagram, groups of 5, 4, and 6 could be represented with a drawn star, or simply a tick mark or X. Your child might show the 4 and 6 being combined to create a group of 10; the answer is correctly represented by 15 objects.

Subtraction

Consider the following equation:

$$5 - 2 = 3$$

Here, 5 is the starting number and 2 is the amount of the change. 3 is the new amount, the difference of the starting number and the change. The Common Core Standards use this example for modeling three primary types of subtraction problems (which you will notice closely resemble the primary types of addition problems).

- **The result is unknown:** Five apples were on the table. I ate 2 apples. How many apples are on the table now? ($5 - 2 = ?$)
- **The change is unknown:** Five apples were on the table. I ate some apples. Then there were 3 apples. How many apples did I eat? ($5 - ? = 3$)
- **The start is unknown:** Some apples were on the table. I ate 2 apples. Then there were 3 apples. How many apples were on the table before? ($? - 2 = 3$)

These formats represent the progression of questions that students will master from kindergarten, first grade, and second grade. It is expected that at the end of first grade, students will be proficient with answering questions where either the result or the change is unknown. Subtraction word problems in first grade are within 20. Effective strategies used for solving problems include using number lines, pictures, and equations.

ESSENTIAL

On occasion your child may be asked to solve the word problem using all three methods (number line, picture, and graph). The philosophy of using all these methods is that representing a problem using more than one strategy can further her understanding even more. On other occasions she may choose one method on her own. When a child is able to choose her own starting point and represent the problem with what makes the most sense to her, math problems are more accessible and understandable.

EXAMPLE 2

Robert's mom left 14 cookies on the table. Robert and his friends ate 5 of them after school. How many cookies are left? Solve the word problem using a picture, a number line, or an equation.

Discussion

For subtraction problems, a diagram could represent 14 cookies with circles, and five of them being eaten represented by crossing off 5 of them. The answer is then represented by the 9 cookies not crossed off. On a number line, the number 14 would be identified as the starting point. Five cookies being eaten would be represented on the number line as moving 5 spaces to the left.

Relating Addition and Subtraction

When students are able to manipulate objects to represent problems, they have a literal experience of adding to some starting quantity and taking away from it. They can often view problems as either subtraction or addition, and relate the two through modeling and by using multiple representations. You will be able to assess your child's understanding of a problem more by observing her working through a problem than just by examining her answer.

EXAMPLE 3

Lily has 15 grapes. Abby has 3 less. How many grapes does Abby have?

Discussion

Lily has the most grapes. They can be represented by the solid bar in this diagram. Abby has 3 less. This means if Abby had 3 more she would have the same amount as Lily.

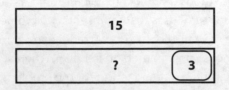

Your child might also write a subtraction or addition equation to find how many grapes Abby has:

$$? = 15 - 3$$

$$? + 3 = 15$$

Addition and Subtraction Equations

In kindergarten students are not required to use equations, though hopefully they will be exposed to them. In first grade, the goal is for students to see the equal sign as a separation between two equivalent quantities. Often students are presented with an expression on the left side of an equal sign and asked to find the answer, such as $12 + 5 = ?$, where the equal sign is almost used to announce "Here comes the answer!"

The Common Core Standards ask for a deeper appreciation of the equal sign, to encourage using equivalent expressions to help students solve problems. For example, students finding the sum of 16 and 4 could break apart the 16 as $10 + 6$ and combine the 6 and 4 to create another group of 10. So they may write $16 + 4 = 10 + 10$, and solve the simpler equation $10 + 10 = 20$. You can help your child appreciate the equal sign more by alternating the side of the equal sign that contains an unknown quantity; for example, using $? = 16 + 4$.

In addition to solving addition and subtraction equations for unknown values, your child will be asked to examine an equation and tell whether the equation represents a true statement or a false statement.

EXAMPLE 4

If the equation is true, circle the word True. If the equation is false, circle the word False.

A.	$16 = 8 + 5 + 3$	True	False
B.	$8 + 7 = 16$	True	False
C.	$5 + 7 = 7 + 5$	True	False
D.	$14 - 3 = 8 + 3$	True	False

Discussion

All equations are true except B. This set of questions asks students to evaluate expressions on both sides of the equal sign and to consider the equal sign as an operator to compare two quantities. Option C demonstrates the *commutative property of addition*. Students can see that the order of addends will not affect the sum. Option D shows the difference on one side of the equation and a sum on the other. This helps students understand the relationship between addition and subtraction ($14 - 3 = 11$ and $11 = 8 + 3$).

Number and Operations in Base Ten

In kindergarten, students began their understanding of place value by seeing the numbers 11 through 19 as one bundle of 10 and some leftover ones. In school, your child will add and subtract multiples of 10 within 100. This practice will help him understand the tens position in place value. For example, he will add 30 to 45 to see that the tens position will change from a 4 to a 7, but that the ones position will remain 5.

Your child will build up his number sense by looking for patterns when comparing numbers within 100, and by thinking of whole numbers between 10 and 100 as groups of tens and ones. This provides a foundation for second grade, when they will bundle groups of hundreds up to 1,000.

In school, your child may use base ten blocks, where a 1 cm cube represents 1, a bar of 10 cubes represents 10, and a 10 × 10 block represents 100. Students will use these to model specific numbers as well as write the numbers represented by the model. For example, the number 34 could be represented as:

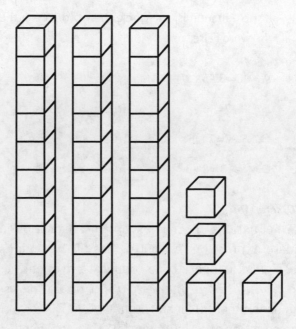

Counting

First graders extend their counting sequence and can start counting from any number between 0 and 120. They can count out loud or count up to 120 objects, and represent their counts with numerals. They can represent numbers with a collection of objects, and group objects in bundles of tens and ones, which establishes the connection between numbers and place value for base ten.

EXAMPLE 5

Fill in the missing numbers.

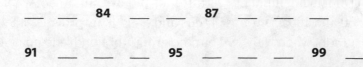

Discussion

When your child completes charts like this, he is practicing many skills at once. First, as he seeks a starting point, he might start with a number that is far less than the first number in the table to get "a running start." As he becomes more proficient with starting to count from any number, the need to get a running start will diminish. Your child has many choices as where to start to answer this question. He may start with 84 and fill in the top row, counting backward from 84 and then forward from 84, or he could see the relationship between the rows and notice that if 91 starts the bottom row, then 81 will start the top row. It is desirable for him to notice that the ones position in each column is the same, and that the tens position in each row is the same, up until the last column.

Place Value

The creators of the Common Core Standards know that students can perform routine addition and subtraction calculations without firmly understanding the concept of place value. However, a deep understanding of place value will provide students with a better understanding of all operations, and provide a better toolbox for problem solving. Students will be able to represent a quantity in different ways, such as numbers and diagrams, and explain the meaning of two-digit numbers. First-grade students will be able to construct and deconstruct numbers and quantities. For example, 15 can be represented by $5+5+5$, or $10+5$. Your child should be able to show this using pictures or numbers.

An important outcome for first grade is for students to be able to thoroughly understand and explain the value of each digit of a two-digit number. They should be able to explain that 25 is $20+5$ and that the 2 represents two bundles of ten, and the 5 represents the ones left over. Students will use words like *bundle* and *compose* when turning groups of ones into a group of tens, and use *unbundle* when they decompose a group of ten into a group of ten ones. This will provide them with a deep understanding for addition with *carrying*, and subtraction with *borrowing*.

Special Cases

Counting objects out loud can make it difficult to visualize groups of tens and ones, but when the quantities are written, or physically arranged in groups of tens, the tens and ones can then be seen. The first two-digit number is 10, which represents exactly one group of ten, with no remaining ones. In kindergarten, students looked at the numbers 11 through 19 as having one group of ten, with between one and nine ones left over. In first grade, students see the numbers 20, 30, 40, 50, 60, 70, 80, and 90 also as having complete groups of ten, with no ones left over.

ALERT

The numbers 11 through 19 can be especially confusing for young learners because they don't follow the naming pattern for the tens and ones used for the other numbers up through 99.

EXAMPLE 6

Cassie has 45 beads. Draw a diagram showing how many groups of 10 she can make.

Discussion

Your child doesn't need to draw the exact object used in a word problem. In this example, beads look like small circles, so there is a connection between the object Cassie has and the objects in the diagram shown.

Tens	Ones
OOOOOOOOOO	
OOOOOOOOOO	
OOOOOOOOOO	
OOOOOOOOOO	OOOOO

Your child may instead represent tens using a long bar (rectangle) and ones using single squares. Your child may understand the problem as a more generic prompt to "draw a picture to show 45," but she is being asked to draw a collection of tens and a collection of ones.

Comparing Two-Digit Numbers

When comparing the quantity of objects in a group in kindergarten, your child could use a one-to-one correspondence between the groups to see which group would have objects left over, as objects were moved from each group one at a time. In first grade, he can use this strategy to remove a single object, or a group of ten objects at a time, from each group he is comparing. He can use the same strategy to compare two numbers.

To show that two groups have the same number of objects, or that two numbers are equal, your child will use the equal sign. He will also use the less than sign (<) and the greater than sign (>) to represent inequalities, relationships that are not equal. Your child can remember which inequality sign to use by having the "point" toward the smaller number, and the open (or bigger) side toward the larger number.

EXAMPLE 7

A. Write three numbers greater than 50 and less than 89.

B. Use <, =, or > to compare each pair of numbers:

31 _____ 39   48 _____ 59   91 _____ 19

C. Which words make this sentence true?

83 _____ 86

is greater than   is less than   is equal to

Discussion

The first question can be answered using any numbers that come after 50 but before 89 as the counting sequence. It's likely that your child will provide an answer that contains some structure or pattern, like even groups of ten (60, 70, 80). Note that 50 and 89 would not be correct answers, because the question states that the numbers must be *greater than* 50 and *less than* 89. When your child answers questions like the last two, she will likely use the counting sequence to find which number comes first; another strategy would be to use a number line and see which number comes first (from the left on the number line) or last.

ESSENTIAL

When talking to your child about comparing a two-digit number to a single-digit number, point this out: The two-digit number will always be greater because it has at least one group of ten, and the single-digit number will always be less than ten. When comparing two-digit numbers, read both numbers from left to right. Talk about how if one number has more groups of ten, it is the greater number; if the tens positions both have the same number, then the ones position will determine which number is greater. If the ones positions also have the same number, the numbers are equal.

Adding Using Tens

Your child will gain proficiency with adding within 100 by understanding place value. He will learn about numbers that make 10 by adding a pair (e.g., 3 and 7, 6 and 4, 9 and 1) and will add numbers by composing tens.

EXAMPLE 8

Find the sum of 45 and 10.

Discussion

When adding 10 or multiples of 10, the ones place does not change. The answer is found by adding the two numbers in the tens place. The answer includes the total groups of 10 (5) and the ones from the first number (5), so the sum is 55. Your child should be proficient with adding and subtracting multiples of 10 whether the problem is presented verbally, horizontally, or vertically like this:

$$
\begin{array}{r}
25 \\
+\,45 \\
\hline
\end{array}
$$

EXAMPLE 9

Add 25 and 46.

Discussion

One strategy for adding 25 and 46 is to see that the 6 in 46 can be decomposed to 5+1, and that the five can be combined to create a group of 10:

$$20+5+40+5+1$$

$$20+10+40+1$$

There are 7 groups of ten, for 70 with one left over, to get 71. Another strategy would be to add the ones places together to get 11, a group of 10 with one left over:

$$20+40+10+1=71$$

EXAMPLE 10

Solve the equation. Explain how you found your answer.

$$77+8=\underline{}$$

When solving the equation 77 + 8, students can see that 77 is 3 away from 80, so they can break the 8 into 3+5, and the 3 can be used to rewrite the equation:

$$80+5=85$$

It's important that your child be able to explain his answers—not just each step he performed, but why he did it and why he knows it must be correct. In this example, a child knows 7 and 3 create a group of 10, and that 8 is the same as 5+3. He still must correctly identify how many groups of 10 there are, and how many ones have not been put into groups of 10.

Subtraction Using Tens

Your child gains proficiency with subtraction within 100 by understanding place value, similar to how he used place value when adding. The Common Core Standards call for first-grade students to gain proficiency for subtracting 10, and multiples of 10, from multiples of 10 in the range of 10–90.

EXAMPLE 11

Jessica had 80¢ and then gave Ivy 30¢. How much does Jessica have left?

Discussion

Your child might write a subtraction equation like this:

$$80¢ - 30¢ = 50¢$$

Or he may use objects or illustrations like 8 bars of ten units to show the original 80¢, and then cross out 3 groups of those ten-unit bars to model taking away 30¢, leaving 5 groups of 10.

EXAMPLE 12

Subtract:

$$\begin{array}{r} 50 \\ -\ 20 \\ \hline \end{array}$$

Discussion

Using vertical subtraction, your child can see that zero is in both ones places. So the ones place in the difference will be 0. The difference between the 5 in the tens place (representing 5 groups of 10) and the 2 in the tens place is 3, so the answer will have a 3 in the tens place representing 3 groups of 10, or 30.

EXAMPLE 13

Solve the equation. Explain how you found your answer.

$$70 - 10 = \underline{\hspace{1cm}}$$

Discussion

Your child can rewrite the equation using a horizontal format, if that makes the problem easier for him to understand. The answer is found by subtracting the corresponding digits, starting with 7 and 1 representing the tens place, with the difference of 6. Because both 70 and 10 do not have any ones, the final answer is 60, representing 6 tens with nothing left over.

With these types of subtraction problems the blank could appear anywhere, such as in the following question:

What number, when subtracted from 70, leaves 60?

$$70 - \underline{\hspace{1cm}} = 60$$

Measurement and Data

First-grade students compare objects by lengths and successfully identify which objects are longest, or shortest, and can order objects based on length. The other measurement standards for the first grade include learning targets for telling time, and organizing and interpreting data.

Measuring Length

Your child is learning to measure lengths indirectly.

EXAMPLE 14

Eddie measured a paper clip that was 2 inches long.

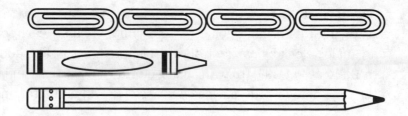

A. How long is Eddie's crayon in inches?

B. How many paperclips long is Eddie's pencil?

C. How much longer is Eddie's pencil than his crayon (in inches or paperclips)?

Discussion

These questions address two standards, measure lengths indirectly and by iterating length units, as students use a paperclip as a unit of measure to find the length of two additional objects. Often the length of the paperclip will not be given, and students answer all questions in terms of paperclip units.

Eddie's crayon is two paperclips long, and each paperclip is 2 inches long. Your child can mark the crayon in terms of inches and then count the number of inches, or count by twos to see that the crayon is 4 inches long. Eddie's pencil can be measured by seeing that each end is in line with the end of the row of four paperclips, so the pencil is four paperclips long. Students have an option of how to specify the length difference between the pencil and the crayon. The pencil is 2 paperclips longer than the crayon. Your child may realize that two paperclips represent 4 inches, and that is the same as the difference between the pencil and crayon.

Time

In first grade your child is formally introduced to telling and writing time using both analog and digital clocks. Your child will learn to measure time using hours and half hours, recognize a clock as the correct tool for measuring hours and minutes, and differentiate between an analog and a digital clock.

Bringing in some geometry, your child will also learn to draw and label an analog clock by partitioning a circle into fourths, and adding the labels 1 to 12. Reading time from an analog clock, your child should be able to write the time using words, such as *four o'clock*, and in digital time, as 4:00.

EXAMPLE 15

Read the analog clock and write the same time in the digital clock.

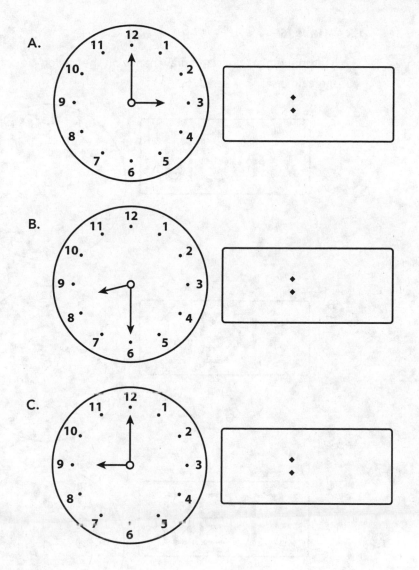

Discussion

The times shown are 3:00, 8:30, and 9:00. A common error is interchanging the hour hand and the minute hand; difficulty also occurs when the minute hand is on the 6 and students do not correctly identify the hour. It can be helpful to show them which way the clock hands move (clockwise) and the clock showing 8:30 has the hour hand past the eight, but before the 9, so the time is before 9:00.

EXAMPLE 16

Read the digital clock and draw hands on the analog clock to show the same time.

A.

B.

C.

Discussion

If your child is having difficulty drawing the hands in the correct position, start by drawing the hour hand. In first grade time questions, the minute hand will either point toward the 12 to represent "on the hour" times, or toward the 6 to represent the half hour. The next learning step is having your child draw the minute hand longer than the hour hand.

Represent and Interpret Data

Your first grader will explore how to represent data. First graders may record data using tally charts, create bar graphs, picture graphs, or use Venn diagrams. The Common Core Standards specify having students organize, represent, and interpret data for up to three categories. Three categories provide a variety, so that one category may have the least of something and one may have the most.

EXAMPLE 17

Mr. Banks keeps the following sports equipment in the gym closet. Draw a bar graph of the sports equipment in the closet. Read your bar graph. How many more hockey sticks are there than baseballs?

Discussion

There are three categories of sports equipment: hockey sticks, baseballs, and soccer balls. Hockey sticks and baseballs are explicitly mentioned in the question. Students begin the data collection process by identifying three different categories of objects in the closet, and then counting the number of objects in each category. The bar graph should be drawn showing the number of objects along the side/vertical axis, and the categories along the bottom/horizontal axis. The height of each bar should represent the number of items in each category. The width of each bar should be approximately the same. A complete graph will have a title and some way to label the axes. The hockey stick bar should represent 8 hockey sticks, the soccer ball bar should represent 3 soccer balls, and the baseball bar should have a height of 6.

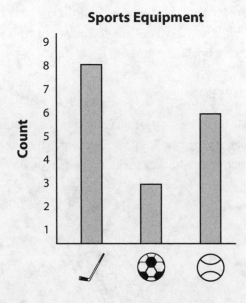

Geometry

In grade 1 students reason with shapes and their attributes. They will describe shapes, such as "is it open like a U, or is it closed?" and they classify closed shapes based on the number of sides or corners. They use two-dimensional and three-dimensional shapes to create larger compound shapes, and partition circles and rectangles into equal shares. By comparing attributes to see how shapes are alike and different, your first grader develops skills that will be used in later grades to understand additional geometric properties, such as congruence and symmetry.

What Makes a Shape a Shape?

The letter L is not a triangle, because it only has two sides. The letter U has three sides, but it also isn't a triangle. The letter T can be seen as having three line segments meeting to create a shape, but it isn't a triangle. So what makes a three-sided figure a triangle? Triangles must have three sides; the sides can only meet at the endpoints; and the shape must be closed.

First graders use attributes such as the number of sides, the number of corners (*vertices*), and whether a shape is opened or closed to classify a shape. They see that shapes have other attributes that are not used to classify the shape—such as color, or size, or how the shape is turned (*oriented*).

EXAMPLE 18

Name each shape. Tell how they are similar and how they are different.

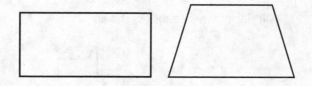

Discussion

When comparing the two shapes, students compare attributes that *describe* the shapes, but don't *define* them, as well as the attributes that do define the shapes. The two shapes are similar—they both have four sides and they both have four corners. Maybe your child will say something about the top and bottom lines being parallel. See if she describes attributes that are unique to trapezoids or rectangles that help define the shapes. She may compare attributes that don't define the shape, such as how they are both filled using white, or no stripes or fill or colors. She might say that both shapes are closed—the ends of each side connect to other sides, and there are no openings. They are both flat (two-dimensional) shapes.

ALERT

Students in first grade are not expected to use vocabulary such as *parallel* and *perpendicular,* but they will find other ways to describe those relationships.

Your child should also note the differences—the angles are not all the same on the trapezoid, but they are in the rectangle. There are two pairs of parallel lines on the rectangle, and not the trapezoid. One takes up more space, so they are not the same size.

In class, your child will be creating shapes using a variety of manipulatives that can include pattern blocks, pentominoes, tangrams, geoboards, tiles, cubes, solids, and other building blocks. She will explore what shapes can be created with other shapes. She will explore attributes of new shapes, such as how many sides a shape has when combined with two or more similar shapes. Her teacher may ask her to build a trapezoid using a fixed collection of pieces. Your child can compare answers and explanations with other children in the class about the shapes they created.

EXAMPLE 19

Will stacked three cubes as shown:

If he repeats this process two more times, which shapes could he make?

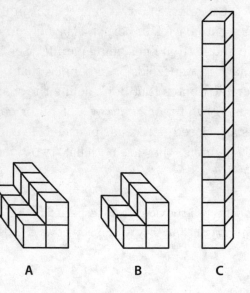

A   B   C

Discussion

In order for Will to make figure A he would have to repeat his shape three more times, so he can't create figure A by repeating his process only two more times. He *can* create figure B. If he used the same number of cubes in figure B but laid them in a line, he would not have enough to create

108

figure C. Students in this example are asked to repeat a pattern, not to create a new pattern with the same number of blocks.

Preparing for Fractions

First-grade students learn to partition circles and rectangles into two and four *equal* shares. When they see shapes partitioned into two or four shares, they determine whether each share is the same size, and if all shares are the same size they can describe the shares as being one half of, or one fourth/quarter of, the whole. They understand that partitioning circles or rectangles into fourths creates smaller shares than when they create halves.

Representing fractions numerically, such as representing one half as $\frac{1}{2}$, is beyond what is specified for first grade in the Common Core Standards. You can assess the progress of your first grader by asking her to perform some of these tasks, and asking her to explain her reasoning.

- Draw a line on a circle or rectangle to show two equal shares.
- Draw lines on a circle or rectangle to show four equal shares.
- Partition a circle or rectangle in half.
- Partition a circle or rectangle into fourths.

The Common Core Standards deliberately chose circles and rectangles because they can be divided into both halves and fourths using horizontal and vertical lines without reorienting the shape, as might be needed with a triangle. Your child may use diagonal lines to also represent equal shares. The focus of the standard is on the shares being *equal* and using the vocabulary *half*, *halves*, *fourth*, *fourths*, and *quarter*.

ALERT

There are many outstanding Internet sites that provide games and practice problems supporting first-grade math. Some websites, such as *www.ixl.com/math/grade-1*, represent fractions numerically, even for first grade. If your child becomes frustrated using online activities, double-check that the math level of the game is a good match for what he is experiencing in the classroom.

EXAMPLE 20

Which pictures represent fourths? Which shape has been partitioned into two equal shares?

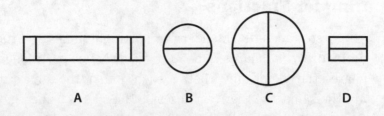

A   B   C   D

Discussion

Rectangle A has been partitioned into four shares, but they are not equal shares; A is not the answer to either question. Picture C represents a circle portioned into fourths, and both B and D have been partitioned into two equal shares.

Helping Your Child Succeed

First grade provides tremendous opportunities for children to take pride in their learning. Your child will continue to learn to count higher and higher and higher. She will be able to perform some mental math problems, and take pride in being able to add more than two numbers at once. Your encouragement will go a long way in developing her appreciation for math, and you can help by showing her how math fits in with her everyday life. When she becomes more proficient with numbers, she will identify patterns and math applications on her own.

Here are some ways you can support your child's schoolwork:

- Stay aware of the topics that are being covered in your child's math class and point out real-world examples.
- Be aware of the strategies that are being developed as your child progresses. Students need to develop multiple problem-solving strategies so they can solve problems that increase in difficulty over time. If your child only focuses on using the simplest strategies to answer a question, without developing a deep understanding of numbers, she will struggle in later grades when the simple strategies no longer solve all of the problems.

- Recognize that the understandings outlined by the Common Core Standards develop over the course of a school year. If your child can't add all of the numbers within 20 midway through the school year, assess whether she can add the numbers within 10 so you can see that she is progressing. Review the math class work she brings home. Address your concerns with the teacher as needed.
- Create your own informal, quick assessments to see what your child can do. For example, draw a rectangle or circle and ask your child to partition it into halves or fourths.

Around the House

Try these games and drills to reinforce what your child is learning in class:

- Play the game Ten More, Ten Less. Select numbers at random from 10 to 90, and ask your child to name the numbers that are 10 more and 10 less. Play this game any time you have a few minutes: in the car, waiting in line at the grocery store. Ask your child to explain his reasoning process.
- Make up addition and subtraction problems that represent real world problems. For example, if there are 24 students in your first grade class, and 22 students in Ms. Allen's class, is there enough room for all of them on a bus that can carry 50 passengers? Just be sure to keep the answers within a range that is appropriate. For example, first-grade students will not have experience with negative numbers, so keep subtraction problems positive.
- Ask time questions. "What time will it be in one hour?" or "What time was it half an hour ago?" Keep in mind that first-grade students work with time in whole-hour and half-hour units.
- Have your child notice the patterns for house numbers, or apartment numbers; odd on one side, even on the other; increasing by twos. Have your child predict the next (or prior) number in the sequence.
- Pose a math question and ask your child to represent the answer in more than one way—maybe a picture, a graph or plot on a number line, an equation, or even the numerical answer spelled out in words.

- Have your child assist with day-to-day counting activities, such as how many glasses to take out for a meal, the number of eggs left in a carton after three are taken out for breakfast, or measure ingredients for cooking.
- Create online bookmarks for websites that have grade-appropriate activities that your child enjoys and that are educational.
- Find online activities that you can adopt for online or offline use. For example, the interactive time-telling activities on the following website can also be used as ways to create questions you can ask your child: *www.shodor.org/interactivate/activities/ClockWise*.

CHAPTER 6

Grade 2

Second graders feel more mature and connected to the world around them as they start working with money and increase their ability to tell time. They will be able to count all the way to one thousand—which is far past the number of objects they would like to count! Grouping objects by tens and hundreds, and adding and subtracting numbers through 1,000 helps them to represent quantities far more easily than counting them all. They will read and write number names, and use their reading skills to work with real-world data.

What Your Second Grader Is Expected to Learn

Second grade is a strong bridge between first grade and third grade. In second grade, your child will find multiple ways for modeling addition and working with repeated groups of the same size. In third grade, she will see multiplication as a fast way to represent repeated addition. Using area models throughout the early grades will help your child with both multiplication and division in later years.

Your child will also:

- Add and subtract numbers through 1,000
- Become fluent with adding and subtracting numbers within 20
- Use strategies for performing calculations using three-digit numbers and for adding up to four two-digit numbers

Additionally, your child will be able to write equations, which she will find helpful for solving one-step and two-step word problems with numbers through 100.

Common Core State Standards: Grade 2

There are four general categories of math skills your child will be learning in second grade, and certain abilities that fall under each of those categories:

OPERATIONS AND ALGEBRAIC THINKING
- Represent and solve problems involving addition and subtraction.
- Add and subtract within 20.
- Work with equal groups of objects to gain foundations for multiplication.

NUMBER AND OPERATIONS IN BASE TEN
- Understand place value.
- Use place value understanding and properties of operations to add and subtract.

MEASUREMENT AND DATA
- Measure and estimate lengths in standard units.
- Relate addition and subtraction to length.
- Work with time and money.
- Represent and interpret data.

GEOMETRY
- Reason with shapes and their attributes.

For the full list of Common Core State Standards for Mathematics, see Appendix A.

What Your Child Should Know Before Second Grade

Children don't all learn at the same pace, so don't interpret this or information from any other sources as a warning that your child is not college- or career-ready. Before entering second grade, your child should know how to count up to 120 starting at any number. If she can't count to ten, she'll struggle in second grade. Your child should also be able to perform addition with two single-digit numbers using mental math. The concepts your child learned in first grade will be built upon in second grade:

- If she can add a single-digit number to a two-digit number, using drawings or models or mental math, then she will be prepared to learn how to add two two-digit numbers and begin to extend her understanding of place value to three-digit numbers.
- If your child can add 10 to, or subtract 10 from, any number from 10 to 90, she is prepared for learning how to do that for any number from 10 to 990. If she can count by tens up to 100 when she enters second grade, she is prepared for learning to count by tens up to 1,000.
- If she is able to compare two two-digit numbers using the $<$, $=$, and $>$ operators, then she is prepared to learn how to compare any combination of two single-digit, two-digit, or three-digit numbers.

Students entering second grade should also know the names of the basic shapes, and be able to identify and use some of their attributes, such as the number of sides. They should be able to sort objects based on length.

Operations and Algebraic Thinking

Students in second grade become proficient adding and subtracting numbers within 100 and are asked to add up to four two-digit numbers. This is a progression from first grade, when they became proficient adding and subtracting within 20. Students become fluent with addition and subtraction by using number patterns and learning strategies to use the structure of numbers, such as finding groups of tens or using numbers that make 100. They prepare for learning multiplication in third grade by adding groups that contain the same number of objects arranged in rectangular arrays.

Solving Problems with Addition and Subtraction

In second grade your child will often use drawing and equations to model and solve addition and subtraction word problems within 100. The word problems he solves may require one or two steps, and you will often see symbols such as blanks, boxes, or circles being used to represent unknown quantities. Your child will also become fluent with addition and subtraction within 20, and know from memory the sum of two single-digit numbers.

EXAMPLE 1

Ms. Holt asked Rachael to share her answer to the following problem with the class. Clara counted 28 clown fish in one tank at the pet store and 16 clown fish in another. How many clown fish did Clara count in all?

Discussion

Here is Rachael's solution: First I showed 28 fish as two groups of 10 with 8 left over:

Then I showed 16 fish as one group of 10 with 6 left over:

卌卌 卌 |

This gave me 3 groups of 10 with 14 left over. I took 10 of the leftover ones and created a new group of 10, then I had four groups of 10 with 4 left over, for a total of 44 clown fish.

卌卌 卌卌 卌卌 卌卌 ||||

In first grade, students answer questions where a starting quantity is known, then there is an unknown change to the quantity, and the result is given; students are asked to find the change. For example: *Clara counted 16 clown fish in a tank at the pet store on Monday, and 24 clown fish in the same tank on Tuesday. How many clown fish were added?* In second grade, your child will be asked to find the starting number where the change and the result are known, as in the following example.

EXAMPLE 2

Emma watched the caretaker at the pet store put 24 new clown fish in a tank. When she counted the fish, there were now 38 fish in the same tank. How many fish were in the tank before the new fish were put in?

Discussion

Here the change is given as 24, and the final result is 38. The starting point is unknown, so this problem can be represented with:

$$\_\_\_\_ + 24 = 38$$

Students can solve this using tape diagrams, number lines, or apply other strategies to find the starting number 14. You can help your child check the answer to make sure it's correct:

$$14 + 24 = 38$$

A tape diagram is an illustration that looks like a section of tape. Tape diagrams help students visualize the relationships between quantities.

In second grade, your child also improves his addition and subtraction fluency within 20 by using mental strategies such as finding ways to use group of tens.

EXAMPLE 3

Fill in the unknown number.

A. ____ + 13 = 20

B. 2 + ____ + 13 = 20

C. 6 + ____ + 4 = 20

D. ____ − 3 = 10

Discussion

In the first three equations, your child learns to find the missing numbers either by asking what number needs to be added to make 20, or by using subtraction to find the difference between 20 and the amount on the other side of the equal sign. The missing number for A is 7 (20 − 13). In equation B, your child might combine 2 and 13 to make 15, then find the missing number is 5 to make 20. She can add 6 + 4 to make 10 in C and find the missing number is 10 to make 20.

A close look at question D shows that 3 is being subtracted from the ones place, and the result has 0 in the ones place, so that 3 must have started in the ones place. The tens place is not being changed, so if the result has a 1 in the tens place then the starting number must also have had a 1 in the tens place; the starting number must have been 13.

Working with Equal Groups

In second grade, your child uses repeating groups that contain the same number of items as a foundation for learning multiplication. He learns about odd and even numbers, and begins counting by twos. Your child begins to see odd-even patterns in number sequences and is able to extend number patterns. He uses *rectangular arrays* to model repeating groups, and express the total by using addition.

Your child also learns that an *even* number is divisible by 2. One way to learn about the concepts of odd and even is to group objects in pairs. When all items within a group can be paired up, the number of items in the group is *even*. When there are "extra" items that can't be paired up, the number of items is *odd*.

EXAMPLE 4

A number is even when you can make groups of 2, and there are none left over. Circle groups of 2 buttons to see if there is an even number of buttons.

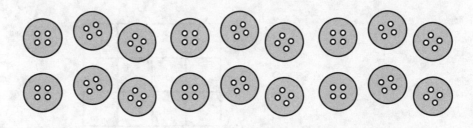

Discussion

Your child will identify that there are 9 groups of 2, for a total of 18. He will be asked to write even numbers as a sum of two numbers. In this example, he would write $9 + 9 = 18$. Numerically, this represents the sum of two equal groups, and the result is an even number. When your child practices with different numbers he will see the pattern that the sum of two even numbers is an even number, and the sum of two odd numbers is an even number.

EXAMPLE 5

Circle the **even** numbers.

1	2	3	4	5	6	7	8	9	10
11	12	13	14	15	16	17	18	19	20

Write the next even number: _____

Write the next odd number: _____

Discussion

In class, your child uses tables like this one, with numbers going up to 100. She'll also begin extending odd-even patterns without using tables. She'll do this by skip-counting (3, 5, 7, or 6, 8, 10, for example). This is an important skill that will be built on in later grades. Skip-counting by 2 leads to skip-counting by 5 and 10, and then more difficult numbers such as 3.

Foundations for Multiplication

Second graders use *rectangular arrays*, such as the following array of 3 rows and 4 columns, to write equations that express totals as the sum of equal amounts.

The sum of this amount is $4 + 4 + 4 = 12$. Arrays are sometimes illustrated as groups of objects that are repeated in rows, like this array of 3 rows of 3 objects, which represents the equation $3 + 3 + 3 = 9$.

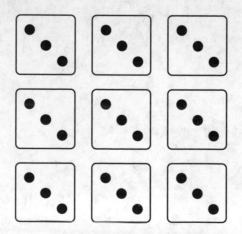

In second grade, your child works with arrays that have up to 5 rows and 5 columns.

120

Number and Operations in Base Ten

In first grade, students use groups of 10 ones to create tens. In second grade, your child uses groups of 10 tens to create hundreds. She adds and subtracts single-, two-, and three-digit numbers within 1,000, and understands that a three-digit number represents a collection of hundreds, tens, and ones. She will also be able to count to 1,000 by ones, fives, tens, and hundreds. Your child learns to read three-digit numbers, compare them, and write them using numerals and number names. She'll write three-digit numbers in standard form, such as $582 = 500 + 80 + 2$.

Hundreds, Tens, and Ones

Your second grader demonstrates her understanding of place value through comparing numbers, arithmetic operations, reading and writing numbers, and counting. Given any three-digit numbers, your child should be able to show that she understands what each numeral means at each position. She can demonstrate this understanding in activities like the following sort-match activity, where she is asked to match numbers, number names, and their expanded forms. In these types of activities, your child matches each number in the center column with an item from each of the other two columns.

▼ **SORT-MATCH HUNDREDS, TENS, AND ONES**

Number Names	Number	Expanded Form
Four hundred twelve	383	200+20+6
Three hundred thirty three	226	300+10+5
Three hundred fifteen	412	300+30+9
Three hundred eighty three	339	300+80+3
Two hundred twenty six	315	300+30+3
Three hundred thirty nine	333	400+10+2

Note: When students write the number names they should not use the word *and*, which should only be used with decimal numbers in later grades.

Comparing Three-Digit Numbers

Your child will be asked to compare two three-digit numbers based on the meanings of the hundreds, tens, and ones places. She will use the >, =, and < symbols to write her results. For example, 333 < 412.

EXAMPLE 6

Fill in the following blanks with <, =, or > to show the relationships between the numbers.

A. 300+30+5 _____ 200+30+2   D. 21 _____ 322

B. 463 _____ 400+40+4   E. 623 _____ 600+20+3

C. 300+3 _____ 400+30+5   F. 439 _____ 433

Discussion

If your child recognizes that A is the expanded form for comparing 335 and 232, she will see that only 300 needs to be compared to 200 to determine the > symbol should be used to show the relationship, and that adding the tens and the ones to the hundreds place is not necessary. Questions such as D help your child see that any number that has a value in the hundreds place will be greater than all two-digit numbers, since the lowest three-digit number is 100 and the highest two-digit number is 99.

If your child is proficient with using expanded forms, she can model or mentally make use of the structure of numbers to make comparisons or perform calculations, and this becomes another strategy in her problem-solving toolbox.

Adding and Subtracting Within 1000

Second-grade students are not expected to use the standard addition and subtraction algorithms or use the terms *carrying* or *borrowing*. Your child is expected to add and subtract numbers within 1000 using other methods that provide her with a deeper understanding for addition and subtraction. Your child applies place value skills, using numbers in expanded form, and properties of operations.

Here are three strategies for solving $568 + 316 = 884$:

Adding using a place value strategy

$568 + 316$

Break both numbers into hundreds, tens, and ones.
5 hundreds plus 3 hundreds equals 8 hundreds.
6 tens plus 1 ten equals 7 tens.
8 ones plus 6 ones equals 14 ones.
Combine 8 hundreds plus 7 tens plus 14 ones: $800 + 70 + 14 = 884$

Adding using the commutative property

$568 + 316$

Write both numbers in expanded form: $500 + 60 + 8 + 300 + 10 + 6$
Reorder the numbers from hundreds, to tens, to ones:
$500 + 300 + 60 + 10 + 8 + 6$
Add 500 and 300 to get 800.
Add 60 to 800 to get 860.
Add 10 to 860 to get 870.
Add 8 to 870 to get 878, then add 6 to get 884.

Adding using a number line

$568 + 316$

Mark the starting point 568 on a number line.
To move forward 316 places, jump 100 three times to get to 868
Jump 10 to 878, then move 6 more places to 884.

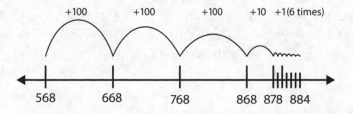

Note: A number line drawn by students does not need to be drawn to scale.

Strategies that work for addition can be applied toward subtraction problems as well. As your child becomes proficient with multiple strategies she may choose one based on the numbers in the problem, or select a strategy she thinks is the best match for the problem. As she gains fluency adding two-digit numbers, she may change her preferred strategy. Here are strategies that can be applied for subtraction, using $568 - 316$ as an example.

Decomposition

$568 - 316$

Break apart both numbers into hundreds, tens, and ones.
8 ones minus 6 ones is 2 ones.
6 tens minus 1 ten is 5 tens.
5 hundreds minus 3 hundreds is 2 hundreds.
What is left is 2 ones plus 5 tens plus 2 hundreds $= 252$.

Think Addition

$568 - 316$

Starting at 316, add 4 to get to 320, then add 80 to get to 400.
Add 168 to go from 400 to 568.
Add $4 + 80 + 168$ using any addition strategy, for example,
$168 + 4$ is 172 and $172 + 80 = 252$

When your child becomes fluent with making 100, she may start with 316 and see right away that adding 84 will bring her to 400, rather than doing it in two steps as was done here. The overall process is to find what is needed to go up to the nearest ten (4 in this case), then what is needed to get to the nearest hundred (80 in this case), and what is needed to get to the final number (168 in this case). This is also the basis for the counting up method often used in stores for returning change to customers.

Measurement and Data

Your child will be busy in second grade with measurement and data skills. Students begin to use standard units of measure, such as feet and yards, and see them as being composed of smaller units. For example, they will see 1 foot as 12 repeated inches. Your child adds, subtracts, compares, and estimates measures of length. Other measurement and data topics include telling time using analog and digital clocks, and solving word problems that involve money. Your child will represent and interpret data in a variety of formats, such as line plots, picture graphs, and bar graphs.

Units of Measure

Rulers, yardsticks, meter sticks, and measuring tapes are all appropriate tools for measuring length. Thermometers are used for measuring temperature, and clocks and calendars for measuring time. When the Standards for Mathematics talk about "using appropriate tools strategically," they are also talking about tools such as *estimation*. Using length is a great way for students to gain estimation skills and apply correct appropriate units. This also gives students practice with assessing the reasonableness of their answers and calculations. Here are some ideas for introducing the idea of estimating to your child:

- Ask your child to estimate how tall he is, and he will likely answer in feet and inches.
- Ask your child to estimate how wide his finger is, and he might answer in inches or centimeters. (Feet, meters, or yards would be less appropriate units of measure.)
- Ask your child to estimate how long a particular car is, and he might answer in feet, yards, or meters. (Inches or centimeters would be less appropriate units of measure.)

There may not be a perfect answer as to the "correct" unit of measure, but some units will be more appropriate than others, and children do not always know what units are customary.

ESSENTIAL

Even adults may not know why a pilot may announce they are flying at 32,000 feet instead of 6 miles. Does putting it in feet make it sound more impressive? The height of tall buildings is also given in feet instead of yards. The distance for a first down in football is 10 yards, but the distance from home plate to first base in baseball is 90 feet. A football field is 100 yards end-zone to end-zone, but a home run is 424 feet to centerfield. Though there can be a discussion of a reasonable measure between feet and inches, there isn't much debate about using inches or miles.

Using Number Lines to Solve Addition and Subtraction Problems

Second-grade students use number lines, which are evenly spaced numbers along straight lines, to model and solve addition and subtraction problems. Your child will probably notice the similarities between number lines and rulers, yard sticks, and meter sticks, which are also evenly spaced.

EXAMPLE 7

Doug measured a pencil that measured 6 inches and a pencil box that measured 11 inches. How much longer is the pencil box than the pencil?

Discussion

Your child might first write the equation:

$$6 + \underline{} = 11$$

He can mark a ruler with the two measurements, then count the difference to see that the pencil box is 5 inches longer than the pencil.

EXAMPLE 8

Aiden's birthday is December 29. If today is December 3, how many more days will Aiden have to wait until his birthday?

Discussion

Make a number line showing the two dates. Count 2 days to December 5, then add 5 to get to December 10, for a total of 7 days. Then count 10 more days to December 20, for a total of 17 days. Finally, add 9 days to get to December 29, for a total of 26 more days until his birthday.

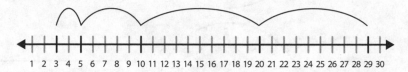

Working with Time

In first grade, students were formally introduced to telling and writing time using both analog and digital clocks, and learned to measure time using hours and half hours. In second grade, your child learns to tell time within 5 minutes, which implies using an analog clock. This ties into other standards, such as counting by fives. Students are often more used to measuring using a straight line such as a ruler or thermometer, so using a clock helps them prepare for other types of measurements that use a dial format.

ALERT

Children in this century can struggle more than children in the last century using dial-type devices. Think of dial-type devices that your child probably hasn't ever used: a rotary telephone, a dial tuner on a television, or a radio dial. House thermostats that control heating and air conditioning were once dial format, but now are being replaced with digital devices.

EXAMPLE 9

Read the analog clock and write the same time in the digital clock.

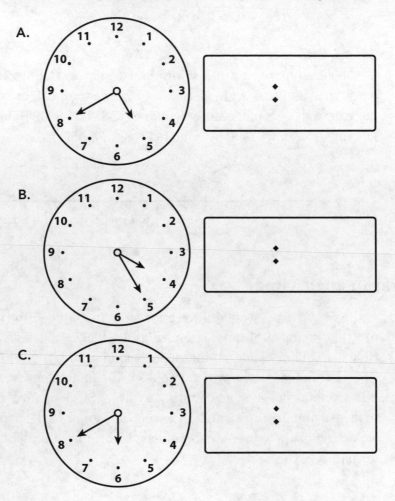

Discussion

You can help your child with telling time by using an analog clock to practice skip-counting by fives up to 60. Help your child understand 15 minutes after the hour as three five-minute periods. Encourage your second grader to use A.M. for times before 12:00 noon, and P.M. for times after 12:00 noon (or exactly 12:00 noon).

EXAMPLE 10

Read the digital clock and draw hands on the analog clock to show the same time.

A.

B.

C.

Discussion

If your child is having difficulty drawing the hands in the correct position, have her start by drawing the (shorter) hour hand. She can determine the position of the (longer) minute hand by counting by fives, starting with the 1 as being the first 5, then moving to the 2 for the second count of 5 (10 minutes), and so on. For second-grade children, representing 5:30 with the hour hand exactly on the 5 rather than halfway between the 5 and the 6 is acceptable.

Working with Money

Second-grade students have generally not been introduced to decimal numbers. (However, some states may have added their own standards introducing money into math curriculum at earlier grades than the Common Core Standards specify.) The money-focused problems that second graders will encounter should use either dollars or cents, but not both. Students should learn some coin equivalencies.

EXAMPLE 11

Kenny buys milk at lunch with four quarters. The milk cost 80¢. List 3 ways that Kenny can receive his change.

Discussion

Kenny paid with $25 + 25 + 25 + 25 = 100$¢. He should get $100 - 80 = 20$¢ change. However, this problem doesn't ask how much is change is given back. It asks for 3 ways to get 20¢ change. This makes the problem more rigorous. Two students can have different answers and they can both be right. Since 20 cents is less than one quarter, no quarters should be involved with an answer. Possible answers include groups of all one type of coin, like 2 dimes, 4 nickels, or 20 pennies. Your child might also come up with coin combinations like 1 dime plus 2 nickels, 1 dime plus 10 pennies, or 1 dime plus 1 nickel plus 5 pennies. There are other possible answers as well.

Working Data

There are two second-grade standards for representing and interpreting data: one to encourage students to measure lengths of objects and to record their results on a *line plot* and the other to represent data with up to four categories. There are online activities that can help children develop graph reading skills. Try *www.ixl.com* and *www.internet4classrooms.com*.

EXAMPLE 12

Lucy asked her classmates if they had ever been to a professional sporting event. She made a graph with the following results.

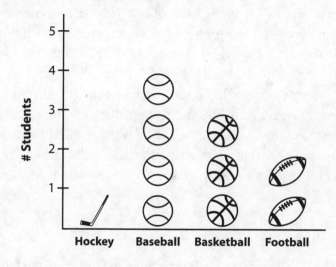

A. How many students have been to either a baseball or basketball game?

B. What sport has the most number of students been to?

C. How many more students have been to a baseball game than a football game?

Discussion

A. 4 students have seen a baseball game, 3 students have been to a basketball game, $3 + 4 = 7$. It is okay for second graders not to consider students that may have said both.

B. Students said they have seen more baseball games than other sports.

C. 4 students have been to a baseball game, 2 have been to a football, $4 - 2 = 2$ more students have been to a baseball game than have been to a football game.

Geometry

In second grade geometry your child is introduced to new shapes, such as the pentagon. He describes shapes with new attributes and may learn new vocabulary such as *angle* or *vertex* to describe a corner. He sees the relationship between the number of sides of a shape and the number of angles. He also continues to use shapes to build, draw, or create other two-dimensional and three-dimensional shapes.

Your child learns to partition a rectangle with lines to create rows and columns and describe the area of the rectangle as the number of same-sized squares. This connects the work he will do with repeating same-sized groups to develop pre-multiplication skills. He applies vocabulary words such as *halves* and *thirds*, and uses words to describe a whole, such as *three thirds*.

EXAMPLE 13

Jason partitioned a rectangle into three rows and four columns. This created equal-sized squares.

The area of the rectangle is the number of equal-sized squares in the rectangle. Count the squares and complete the sentence:

The area of the rectangle is _____ square units.

Discussion

In second grade, students are asked to partition a rectangle into equal-sized squares and come to understand area as the total number of equal-sized squares. This method is used in later grades to find the area of more irregularly shaped areas such as a room or a floor plan.

EXAMPLE 14

Use the images to answer the questions. There may be more than one picture used to answer each question.

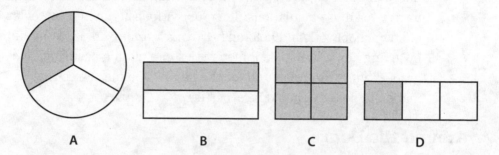

A   B   C   D

1. Which pictures show one half of the shape shaded?

2. Which pictures show one third of the shape shaded?

3. Which pictures show two equal parts with one of the parts shaded?

4. Which pictures show four fourths of the shape shaded?

Discussion

Picture B shows one half of the shape shaded. Pictures A and D represent one third. In both cases the whole is partitioned into three pieces, with one third being shaded. Students often see similar partitions like this when the shapes are different, but can need practice seeing that partitions of the same whole do not have the same shape. Picture B shows two equal parts with one half of the parts shaded. This is a slight variation of the first question. Picture C shows four fourths of the shape shaded.

When students are asked to partition wholes into same-sized pieces they may use diagonals or find other creative methods. For example, here are two additional ways the rectangle in the previous example can be partitioned into equal halves.

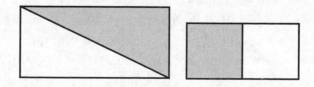

Helping Your Child Succeed

Second grade should be a fun time for your child to learn new things, like improving her ability to tell time and being able to use mental math to perform two-digit calculations. Keep up with assignments by reviewing your child's schoolwork and helping her to stay organized. Keep her interests up by giving her a yardstick or meter stick or a cloth measuring tape. (Second-grade fingers are no match for the metal tape measures that roll up automatically, so they should be avoided.)

Around the House

When you and your child have a free moment, try to work in a math question in the form of a game that can take less than one minute to play. This could be time at a stoplight on the way to school, between commercials on TV, or before bedtime. Here are some ideas to make learning a bit more fun:

- You could play the More or Less game: Select a random number from 100 to 900 and ask your child to name the number that is 10 or 100 more or less than the number you select. You can specify add or subtract, and vary the questions between 10 and 100. Keeping the numbers you select between 100 and 900 will keep her answers between 0 and 1000.
- In the Think Addition game, you select a number less than 1000 and ask your child what number is needed to go up to the nearest 10, or up to the next 100. This will help her develop both addition and subtraction fluency. You can also adapt this by posing a question in this format: "You have $46. How much do you need to have $70?" This will help her develop a two-step process—adding 4 to get to 50, then adding 20 more to get to 70. These questions are great for developing mental math skills, and don't require paper and pencil. It's fine if your child uses fingers, however!
- Help your child learn to tell time by asking her what time it is, or what time it will be in five or ten minutes. Help her get a sense of elapsed time. For example, ask her to estimate how long it will take to get to school (or the library or park). Have her time the ride.

- Ask your child to model odd and even numbers using pennies, buttons, or other manipulatives. Select a number from 1 to 20 and ask her to count out that many items. Then have her create groups of two to decide if the number is odd or even. Ask if there can ever be more than one left over, and have her show why there can't be. When she is proficient with odd and even numbers, she will explain that if there is more than one left over, then another pair can be made. (You can extend this game to include numbers higher than 20. Once she sees the pattern that even numbers end with 0, 2, 4, 6, or 8 and odd numbers end with 1, 3, 5, 7, or 9, you can ask questions about higher numbers beyond the number of items she can model with.)

- Have your child measure objects around the house, selecting tools that are most appropriate for the job. For example, have her measure the dimensions of a room or table using a tape measure, or the height of a chair using a yardstick or meter stick. Ask her to measure toys, books, CDs or DVDs, and smaller objects using a ruler. To show how units differ, have your child measure the same object, once in inches and once in feet, or once in centimeters and once in meters. Ask her to describe how the two units are related.

- After your child is familiar with measuring objects, ask her to first make estimates of the length of objects, and then have her check the estimates by measuring the objects.

Exercises

1. Tommy found 13 frogs in the pond, and Zach found 20. How many total frogs did the boys find?

2. Write an equation to express the total number of squares in the following grid:

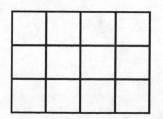

3. There are 43 students in second grade at Johnson Elementary School, and 36 students in first grade. How many more students are in the second grade?

_____ – _____ = _____

4. Jazzy has 10 fewer hair ties than Helen. Helen and Ava both have 60 hair ties. Becca has the fewest hair ties. She has 40. How many hair ties do the four girls have all together?

5. Solve: 675 + 126. Explain your answer.

6. Brad has two quarters, two dimes, and two pennies. How many cents does Brad have?

CHAPTER 7

Grade 3

In terms of math, third grade is an important year filled with many new concepts. It's also the year that your child may encounter his first Common Core assessment test. Luckily, third grade does a great job building upon the experiences from second grade; for example, the skill of counting equal-sized groups will grow to form an understanding of multiplication and a reference point for introducing division. Your child will develop an understanding for multiplication and division using numbers up to 100. He will also use unit fractions, such as $\frac{1}{2}$, $\frac{1}{3}$, and $\frac{1}{4}$, to begin to understand fractions as parts of a whole. He will also learn more about geometry by expanding his understanding of two-dimensional shapes.

What Your Third Grader Is Expected to Learn

Mathematically, the biggest gains for your third grader are centered around multiplication, division, and fractions. Connections will be made to material learned in second grade, and you will see your child gain proficiency with his second-grade skills as they are applied in third grade.

In second grade, your child partitioned areas or objects in halves, quarters, or thirds. In third grade, he will represent these fractions numerically, such as $\frac{1}{2}$ or $\frac{1}{3}$. Your child will also learn to compare fractions, plot them on a number line, and read them from a number line.

In second grade, he learned to view 100 as ten groups of ten items, and learned that three-digit place value represents some number of hundreds, tens, and ones. In third grade he will use place value to add and subtract numbers up to 100, multiply single-digit numbers, and divide numbers up to 100. Place value skills will include rounding and multiplying numbers up to 90 by ten.

QUESTION

Does Common Core make third graders deal with remainders?
No. Division problems are expected to result with a whole number. Students are not expected to experience division problems with fractions as remainders. A division answer that doesn't result in a positive whole number is a signal that a problem-solving error has occurred.

Your child's first exposure to a Common Core assessment is likely to take place in third grade. Most states will administer either the PARCC assessment or assessments from the Smarter Balanced Consortium. States that have adopted the Common Core State Standards may choose to develop their own test, based on their adaptation of the standards. Problem-solving strategies learned in third grade may include reviewing possible solutions, identifying ones that are possible, and others that are not feasible.

During this year, your child will gain experience taking online assessments, experience online interactive lessons and games, and learn technology tools for modeling and solving problems. He will experience material associated with each of the core Standards for Mathematical Practice and be assessed not just on the answers he produces, but also on his ability to demonstrate and explain his problem-solving process.

ALERT

You may find that your child exceeds the learning outlined by the standards. For example, he may extend his understanding of multiplication beyond numbers 1 to 100, such as adding 4 groups of 51 to get 204. His ability to tell time, manipulate money, or perform area and volume calculation may exceed what he is tested on in multistate assessments.

Common Core State Standards: Grade 3

There are five general categories of math skills your child will be learning in third grade and certain abilities that fall under each of those categories:

OPERATIONS AND ALGEBRAIC THINKING
- Represent and solve problems involving multiplication and division.
- Understand properties of multiplication and the relationship between multiplication and division.
- Multiply and divide within 100.
- Solve problems involving the four operations (addition, subtraction, multiplication, and division) and identify and explain patterns in arithmetic.

NUMBER AND OPERATIONS IN BASE TEN
- Use place value understanding and properties of operations to perform multi-digit arithmetic.

NUMBER AND OPERATIONS—FRACTIONS
- Develop understanding of fractions as numbers.

MEASUREMENT AND DATA
- Solve problems involving measurement and estimation of intervals of time, liquid volumes, and masses of objects. Represent and interpret data.

- Geometric measurement: understand concepts of area and relate area to multiplication and to addition.
- Geometric measurement: recognize perimeter as an attribute of plane figures and distinguish between linear and area measures.

GEOMETRY
- Reason with shapes and their attributes.

For the full list of Common Core State Standards for Mathematics, see Appendix A.

What Your Child Should Know Before Third Grade

Students entering third grade should know how to add two single-digit numbers fluently and be able to use mental math strategies to add and subtract numbers within 20. They should be able to perform calculations using addition and subtraction to solve problems within 100; and they should be able to count to 1,000 by 1s, 5s, 10s, and 100s. They should understand place value with three-digit numbers as hundreds, tens, and ones. They should be able to use their understanding of place value to compare two three-digit numbers using <, >, and =. They should be able to mentally add 10 or 100 to any number from 1 to 900 and subtract 10 or 100 from a given number from 100 to 900. They should have experience using rectangular arrays, be able to partition a rectangle into rows and columns of same-sized squares, and be able to count the total number of squares in an area.

Operations

In third grade your child will become fluent with multiplying two single-digit numbers. In other words, she will learn the times table. She'll use models that represent a group of items being repeated to visually demonstrate the connection between addition and multiplication. As she multiplies two single-digit numbers, she'll also develop fluency with dividing numbers within 100 by single-digit divisors. She will also see that repeated subtraction leads to division. Your child will learn the connection between multiplication and division, and use all four operations (addition, subtraction, multiplication, division) to solve word problems. She will learn the properties of operations.

QUESTION

Multiplication and Division

In second grade, students should see the following illustration as 3 groups of 10, for a total of 30.

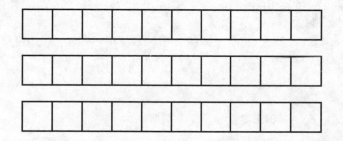

In third grade, students interpret this type of model as $3 \times 10 = 30$. They begin to see multiplication problems as *some number of groups* times *some group size* as the total. The result found by multiplying is called a *product*. Your child uses visual models to see the product of two whole numbers as the total of some number of groups (in this case 3 groups) times some group size (in this case 10). She also uses visual models to develop her understanding of division.

EXAMPLE 1

Mrs. Miller has 24 students in her gym class. She wants to divide the students into two groups. How many students will be in each group?

Discussion

This problem requires students to find the unknown group size, given the total and the number of groups (2). The two groups Mrs. Miller needs can be represented by a row of boxes. Because Mrs. Miller wants the two groups to have the same number of students, count (or label) the students as they are put into two groups:

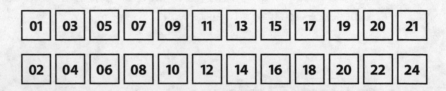

24 students ÷ 2 groups = group size

The model shows the process of dividing 24 students into two groups with 12 students in each group.

2 groups × ? group size = 24 students

$$2 \times 12 = 24$$

EXAMPLE 2

Mrs. Miller has 24 students in her gym class. She wants to divide them into teams of 4 students. How many teams will Mrs. Miller have?

Discussion

This problem requires your child to find an unknown number of groups, given the total and the group size. She can represent one group by the following.

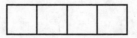

Repeatedly adding one group of 4 students until there are a total of 24 students shows that Mrs. Miller will have 6 teams of 4.

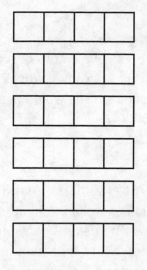

24 students ÷ group size (4) = number of groups

24 ÷ 4 = 6. ? groups × group size (4) = 24 students

6 × 4 = 24

EXAMPLE 3

Which expression is represented by the following model?

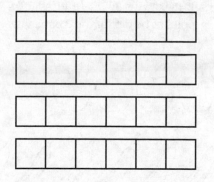

A. 4 × 6

B. 6 ÷ 4

C. 24 × 4

D. 6 ÷ 24

Discussion

There are 6 objects in each row, so the group size is 6. The model shows 4 groups of 6, or 4×6. Your child may have counted, or calculated, that there are 24 blocks in total, and that they have been divided into four rows, but $24 \div 4$ is not one of the answer choices. She would need to think about another way $24 \div 4$ can be stated (4×6).

Multistep Word Problems

Third graders are expected to answer word problems that require more than one calculation step. Your child is developing multiplication and division skills, so the word problems will be using two single-digit numbers to keep the products under 100. Models that show multiple groups will be showing groups that are all the same size. The models help to restrict multiplication and division problems to only using whole numbers. Equations are also a good way to model a problem, using a symbol to represent an unknown number.

EXAMPLE 4

In a computer game about farms, players earn points by collecting apples and bananas. The following picture shows the apples and bananas that Timmy collected while he played the game. Timmy earned the same number of points for the 6 apples as he did for the 8 bananas.

If Timmy earned 24 points for the apples, how many points did he earn for each banana? Show your work.

Discussion

With an open-ended question like this, students are expected to demonstrate their understanding of multiplication and division; a student who simply writes "3" will not earn any points on a standardized test. Though arriving at the correct answer, the student wouldn't have demonstrated a proficient use of either multiplication or division. Student responses will vary, but should answer the question "How many points are awarded for each banana?"

ALERT

When students show their work, it should be neat and precise. With multi-step processes, it's important to be clear about what is known, and what is unknown. Keeping work neat and organized will reduce calculation errors and calculations with incorrect values.

Your child's work should show that because Timmy earned the same number of points for the apples as the bananas, and since he earned 24 points for the apples, he must have also earned 24 points for all of the bananas. He could then show this as a missing factor problem, or use other models to arrive at the solution. For example:

$$24 \div 8 = 3$$

$$8 \times ? = 24$$

He could see this as needing 8 groups, one for each banana, and then use tally marks to find out what the value must be for each group.

More Modeling

The Common Core Standards impart deep understanding for operations by using visual models. When students are shown the standard algorithms, they won't just be repeating steps procedurally; they will understand each step and use the structure of numbers and properties of operations to find creative solutions and gain deeper problem-solving skills.

EXAMPLE 5

Find the product of 17×4.

Discussion

Think about this problem in terms of how it can be modeled. 17×4 is 17 groups with a group size of 4. A group size of 4 can be modeled with the following:

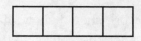

If you visualize 17 as $10 + 7$, you'll need to have 17 groups: 10 groups, then 7 more groups, as shown here.

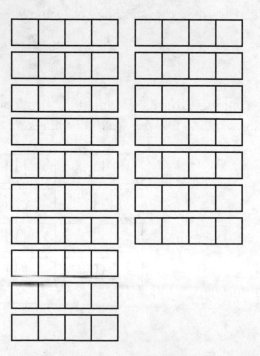

$$10 \text{ groups of 4 is } 10 \times 4 = 40$$

$$7 \text{ groups of 4 is } 7 \times 4 = 28$$

$$17 \text{ groups of 4 is } 17 \times 4 = 68$$

This is an example of using the *distributive property*.

Finding the total of 17 groups of 4 or (17×4) is the same as finding the total of 10 groups of 4 plus another 7 groups of 4.

$$17 \times 4 = (10 \times 4) + (7 \times 4) = 40 + 28 = 68$$

FACT

The *distributive property* states that multiplying a group of numbers added together is the same as doing each multiplication separately. So $2 \times (3+4)$ is the same as $(2 \times 3) + (2 \times 4)$. So the 2 can be "distributed" across the $3+4$ into 2 times 3 and 2 times 4.

EXAMPLE 6

What number goes in the blank to make the number sentence true?

$$13 \times 3 = (? \times 3) + (3 \times 3)$$

A. 10

C. 17

B. 13

D. 26

Discussion

Think about what a model of this would look like: 13×3 is 13 groups of 3. Having 13 groups of 3 is the same as having 3 groups of 3 plus some additional (unknown number) groups of 3. So how many more groups of 3 would you need? $13 = 10 + 3$, so 13 groups of 3 is the same as having 10 groups of 3 plus 3 groups of 3.

$$13 \times 3 = (10 \times 3) + (3 \times 3)$$

$$13 \times 3 = (10 \times 3) + (3 \times 3)$$

10 groups of 3 is 30, and 3 groups of 3 is 9, so the total is 39.

Please remember that your child is not expected to master material the first time he sees it. Being able to decompose 39 into $30 + 9$ will provide a better understanding for future math than memorizing the multiplication chart. You will be amazed with the mental math your child can do when he utilizes the many structures that numbers have and understands the properties of operations!

Relating Division to Multiplication

Before students are taught the standard algorithms for division they experience division as having to find the *unknown factor*. Consider: If you have 30 cookies to divide among 15 people, you would calculate $30 \div 15 = 2$. Think of this as a multiplication problem with a missing factor: There are 15 groups and total of 30, but how many are in each group? $30 = 15 \times$ _____.

The model used so far for multiplication has been:

$$\textbf{number of groups} \times \textbf{group size} = \text{total}$$

Your child should understand these division problems:

$\text{Total} \div \textbf{number of groups} = \textbf{group size}$ as a missing factors problem:
$$\textbf{number of groups} \times ? = \text{total}$$

$\text{Total} \div \textbf{group size} = \textbf{number of groups}$ as a missing factors problem:
$$? \times \textbf{group size} = \text{total}$$

EXAMPLE 7

Erica separated 45 quarters from her collection into 5 equal stacks. What number sentence could be used to determine the number of quarters in each stack?

A. $5 \div ? = 45$ **C.** $? + 5 = 45$

B. $5 \times ? = 45$ **D.** $? \div 5 = 45$

Discussion

The first step of this word problem is to understand what is given and what your child is asked to find. The total is given to be 45. The 5 equal stacks represent the groups, so your child is being asked to find the group size.

$$5 \text{ groups} \times \text{some group size} = 45$$

This is represented by choice B: $5 \times ? = 45$. The number 45 can be viewed as 5 groups of 9 quarters.

EXAMPLE 8

What number sentence could be used to determine the value of $56 \div 8$?

A. $8 \times ? = 56$

B. $8 + ? = 56$

C. $8 \times 56 = ?$

D. $8 + 72 = ?$

Discussion

You are dividing the total, which is 56. You are looking for two factors of 56, and one of them is 8. There isn't any information about whether the 8 represents the number of groups or a group's size. So if you want to divide 56 into 8 groups, how many would be in each group?

$$8 \times ? = 56$$

If you want to divide 56 so that each group contains 8 objects, how many groups can be created?

$$? \times 8 = 56$$

The commutative property of operations says that the order of the factors does not matter, so $7 \times 8 = 56$ and $8 \times 7 = 56$. But the question isn't asking for the answer of 7. It is asking which number sentence could be used to find the value. Notice also that there are no division symbols in any of the answer choices. This problem is not asking what 56 divided by 8 is; it is asking if your child understands division as an unknown-factor problem.

Solving Word Problems

Your third grader will use the four operations to solve one-step and two-step word problems. In third grade, sometimes more than one operation is needed to solve a particular problem. Students will use multiple operations to describe patterns and sequences.

ESSENTIAL

Students will see many ways to represent unknown values such as _____ or ?. Letters may also be used to represent unknown values, but students won't refer to them as *variables* until sixth grade.

EXAMPLE 9

Tommy has 42 basketball cards and 18 baseball cards. Tommy used all of his cards to fill a 10-page notebook. He put the same number of cards on each page. How many cards did he put on each page? Check your answer.

Discussion

The first step in solving this problem is recognizing that the number of basketball cards and the number of baseball cards should be added together to get the total number of cards. Tommy has a total of $42 + 18 = 60$ cards.

The second step is to divide the sum by 10, because there are 10 pages. Students may see 10 as the number of groups, and they need to solve for the size of the group.

Here is a possible model for this problem:

The number of groups $\times$ the group size = total number of cards

The number of groups is the number of pages, 10. Let c be the group size, which is the number of cards on one page.

$$10 \times c = 60$$

The number that makes this number sentence true is 6, so there are 6 cards on each page.

Here's how you check the work: If there are 10 pages, each with 6 cards, then $10 \times 6 = 60$ cards can fit in the notebook.

Numbers and Operations in Base Ten

In third grade, students are expected to become proficient adding and subtracting three-digit numbers using their understanding of place value for ones, tens, and hundreds. Your child begins to see the arithmetic patterns of addition, such as $7+8$ is the same as $8+7$. He becomes more familiar with vocabulary words that may suggest an operation. For example, hints that addition may be needed for solving a problem may be found with keywords or phrases such as *in total*, *altogether*, *increased by*, *combined*, or *more*. Hints that subtraction may be needed are *the difference*, *who has more*, *by how many*, *minus*, or *decreased by*.

Your third grader begins to use number sentences and equations to model problems. He uses place value strategies and the properties of operations to add and subtract within 1,000 and to round whole numbers to the nearest 10 or 100. He also extends his ability to multiply single-digit numbers by 10 to multiply single-digit numbers by multiples of 10.

Multiplying by Multiples of Ten

A single digit specifies a number with 0 tens and some number of ones. When a single-digit number, such as 8, is multiplied by 10, it shifts one place value position to the left. The number 8 originally represented the number of ones, and now it represents the number of tens. This can be represented using a place value table:

$$8 = \begin{array}{|c|c|} \hline 10\text{s} & 1\text{s} \\ \hline 0 & 8 \\ \hline \end{array} \qquad 80 = \begin{array}{|c|c|} \hline 10\text{s} & 1\text{s} \\ \hline 8 & 0 \\ \hline \end{array}$$

8 ones multiplied by 10 becomes 8 tens, 0 ones.

When a two-digit number is multiplied by 10, both numbers are shifted one position to the left. The following place value table represents $84 \times 10 = 840$.

$$84 = \begin{array}{|c|c|c|} \hline 100\text{s} & 10\text{s} & 1\text{s} \\ \hline 0 & 8 & 4 \\ \hline \end{array} \qquad 840 = \begin{array}{|c|c|c|} \hline 100\text{s} & 10\text{s} & 1\text{s} \\ \hline 8 & 4 & 0 \\ \hline \end{array}$$

8 tens 4 ones multiplied by 10 becomes 8 hundreds, 4 tens, 0 ones

Your third-grade student is learning to multiply two single-digit numbers, and to multiply any single-digit number by 10. Using place value, this can be extended so students can multiply any single-digit number by factors of 10. For example, 7×60 can be seen as $7 \times 6 \times 10$, which becomes $42 \times 10 = 420$.

EXAMPLE 10

Multiply:

A. 10×4

B. 40×4

C. 8×80

Discussion

Students will learn 10×4, either from the multiplication tables, through thinking of this as four groups of 10, or by counting $10 + 10 + 10 + 10 = 40$. Throughout the Common Core Standards, the intention is to stress the importance of applying the properties of operations. One of the properties of multiplication to consider here is that the order in which two numbers are multiplied results in the same product, and that if 40 is decomposed to 10×4, then question B becomes $4 \times 10 \times 4$ or $4 \times 4 \times 10$ which students can solve as $16 \times 10 = 160$. Similarly, 8×80 becomes $8 \times 8 \times 10 = 64 \times 10 = 640$. This leverages students' abilities to multiply two single-digit numbers, and multiply a number by 10.

Adding and Subtracting within 1,000

When your child first began to add three single-digit numbers together, for example $3 + 6 + 7$, she used the structure of numbers and the properties of operations to look for pairs of numbers that would add up to 10, and perform the calculations after rearranging the numbers: $3 + 7 + 6 = 16$. These same strategies will help your child to add two-digit and three-digit numbers by looking at the structure of numbers to form tens, hundreds, and thousands.

EXAMPLE 11

There are 165 third-grade students at Beaver Elementary School, and 143 third-grade students at Carpenter Elementary School. The auditorium at the Town Playhouse can hold 300 people. Can both schools send all of their third-grade students to see a play at the same time? Explain your answer.

Discussion

First, find the total number of third-grade students:

$$165 + 143$$

Think of 165 as $100 + 60 + 5$, or 1 hundred 6 tens and 5 ones. Think of 143 as $100 + 40 + 4$, or 1 hundred 4 tens and 3 ones. Adding these by place value results in:

$$2 \text{ hundreds, } 10 \text{ tens, and } 8 \text{ ones}$$

10 tens can be regrouped as 1 hundred, so the sum becomes:

$$3 \text{ hundreds, } 8 \text{ ones} = 308$$

Since there are 308 students, which is more than 300, there are not enough seats for the students from both schools to see the play.

This example could also be viewed as a subtraction problem. If all of the students from Beaver Elementary School were in the auditorium, would there be enough seats for the students at Carpenter?

$$300 - 165 = \underline{\quad}$$

If 5 students from Carpenter were added there would be 170. If 30 more students were added from Carpenter, there would now be 200. If 100 more students were added from Carpenter, the auditorium would be full with 300 students. The total students from Carpenter were $5 + 30 + 100 = 135$, which is not all of the students from Carpenter, so they would not all fit.

EXAMPLE 12

What unknown number makes this equation true?

$$505 - 256 = \underline{\quad}$$

Discussion

Using the *count-up* method, think about what number, when added to 256, will total 505.

> Start with 256 and add 4 to get 260.
> Add 40 to get the total to 300.
> Add 205 to get to 505.
> $4 + 40 + 205$ was added to 256 to get to 505
> $4 + 40 + 205 = 249$

This means $505 - 256 = 249$, and was modeled by finding that $256 + 249 = 505$.

EXAMPLE 13

The difference between two numbers is 70. If the bigger number is 190, find the smaller number.

Discussion

One student could think about this as "if the difference between the big number and the small number is 70, then I have to add 70 to the smaller number to be equal to the larger number," and write $70 + \_\_\_ = 190$. Another student who uses a more literal translation of the word problem could read the first sentence, then write:

$$\_\_\_ - \_\_\_ = 70$$

to show that the difference between two numbers is 70. When she reads the second sentence, she could replace the bigger number with 190:

$$\underline{190} - \_\_\_ = 70$$

Either method will lead to finding the smaller number, which is 120.

Rounding

Being able to round numbers to ten or hundreds is a great tool for using estimations. Students can use estimations to check the reasonableness of their answers or even to monitor their own progress while solving multistep problems. Students use rounding to help explain their answers, and when questions are asking for approximate answers.

EXAMPLE 14

Mrs. Sutton has 18 students in her Math Club. She wants to give everyone 5 index cards, but she only has 100 cards. Does she have enough cards?

Discussion

Use rounding to estimate the answer. Round 18 students to 20. Does she have enough cards to give 20 students 5 cards?

$$\text{Solve: } 20 \times 5 = \underline{\quad}$$

Using place value strategies and the properties of operation, this can become

$$2 \times 10 \times 5 = 2 \times 5 \times 10, \text{ or } 10 \times 10 = 100$$

Because 18 was rounded up to 20, which increased the number of students, and she has enough for 20 students, then she has enough for 18 students.

Number and Operations—Fractions

Fractions aren't really that hard to understand. When you use a fraction, like $\frac{1}{2}$ or $\frac{1}{3}$, you give numeric meaning to a single piece of something that has been split up. In other words, fractions represent one part of a whole. It's a good idea to talk about the idea of fractions often when you talk to your child. Ask your child to split a group of items evenly or figure out how many people could share a candy bar that has been cut into quarters. Talk about what's better—having an eighth of a pizza, or having a sixth. Look for examples of fractions in everyday life.

Unit Fractions

A *unit fraction* is a special case fraction where the top number (the numerator) is 1, and the bottom number (the denominator) is a positive whole number. In third grade, your child learns about the concept of fractions being "parts of a whole" as they work with $\frac{1}{2}$, $\frac{1}{3}$, and $\frac{1}{4}$. Illustrations like the following show how a whole is broken up into fractions:

This year, your child also learns about fractions as they relate to the relative size of the whole. For example, $\frac{1}{3}$ of a larger whole can be bigger than $\frac{1}{2}$ of a smaller whole. Third graders develop strategies to compare unit fractions by comparing equal numerators or denominators.

EXAMPLE 15

Lisa has a flower garden partitioned into equal pieces. The shaded part of the diagram shows where Lisa planted roses.

What fraction of the garden is planted with roses?

A. $\frac{1}{5}$ C. $\frac{1}{10}$

B. $\frac{1}{2}$ D. $\frac{1}{9}$

Discussion

The key to solving the problem is to look at the number of total equal parts, 10 in this diagram. The fraction that matches the diagram will have a denominator of ten. Choice C represents a whole partitioned into 10 equal parts, with 1 of the parts shaded.

EXAMPLE 16

Plot each of the following fractions on the following number line.

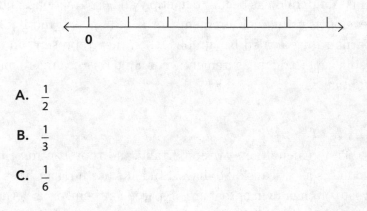

A. $\frac{1}{2}$

B. $\frac{1}{3}$

C. $\frac{1}{6}$

Discussion

Students should see that they can partition the same whole into different, same-sized lengths. After plotting, they should see that the length of $\frac{1}{6}$ is less than the length $\frac{1}{3}$, which is less than the length $\frac{1}{2}$, and that the more pieces a whole is partitioned into, the smaller each unit becomes.

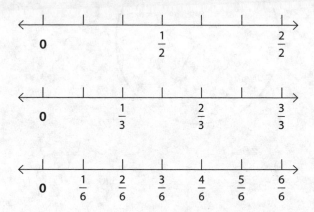

Measurement and Data

Third graders improve precision with telling time, and build time estimation and calculation skills. Your child will be able to tell time using an analog clock accurate to the nearest minute. She'll also use number line diagrams or other visual models to help solve time addition and subtraction word problems. Your child has begun to multiply and use fractions, and these skills are developed using measurements for length, mass, and liquid volumes. She learns to use scaled bar graphs when they represent and interpret data. Using geometric measurement, your child relates area to multiplication and addition.

Time

Time calculations are important, and representing time using visual models is a critical skill to have. Students use number lines that either they create on their own or that are prepared for them, or use other visual models to record time and timed events.

EXAMPLE 17

What time is shown on the clock?

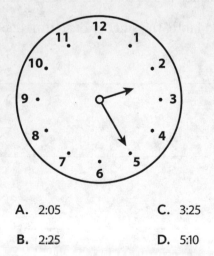

A. 2:05   C. 3:25

B. 2:25   D. 5:10

Discussion

Answering correctly requires identifying the position of the hour hand as representing 2 o'clock and the position of the minute hand as representing 5×5 minutes after the hour, so the correct time is 2:25.

Your child could find several ways to compute elapsed times, total times, or solve word problems that ask for calculating starting and ending times. Visual models include clock diagrams representing times and number lines (time lines). Students counting total elapsed minutes often forget that there are 60 minutes in one hour, not 100.

EXAMPLE 18

Rita answered the following question for homework.

Paulo gets home from school at 3:10. He will work for 1 hour in the yard before walking to the store to get some milk. If Paulo spends 15 minutes walking each way to the store and spends 10 minutes in the store, will he get home in time to watch the 5:00 news? Explain your answer.

Discussion

Here is Rita's solution: Using a number line as a time line, I first marked 3:10 because that is when Paulo gets home.

When he finishes in the yard it is 4:10. When he gets to the store it is 4:25. He is in the store for 10 minutes, leaves at 4:35, and is home at 4:50. He will be home 10 minutes before the 5:00 news starts.

Rita's time line doesn't use a consistent spacing on the number line. The distance from 3:10 to 4:10 represents 1 hour, and the same distance is used to indicate the ten minutes between 4:25 to 4:35. This could make it confusing for someone else to read, but it supports Rita's explanation.

Geometric Measurement: Area

Your child is learning to recognize *area* as an attribute of two-dimensional regions, building on the attributes of length and width learned in previous grades. He measures the area of a region by finding the total number of same-sized units required to cover the shape without gaps or overlaps.

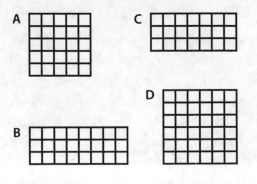

This diagram shows 30 equal-sized squares filling an area. Each square could represent one square inch, one square centimeter, or even larger square units. In class, your child is learning to decompose other shapes into equal-sized pieces, and to connect finding a shape's area to multiplication.

EXAMPLE 19

A garden has an area of 24 square meters. Which shape could represent the garden?

KEY

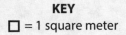 = 1 square meter

A   C

B   D

Discussion

The choices for this question are intentionally rectangular. It is hoped that students will use repeated addition or multiplication to find the total number of squares in each area. For example, when students are counting the squares of area A, they could count 5 squares across, then count down by 5 to find that area A has an area equal to 25 square units. They might see that there are 3 rows of 7 squares in choice C and use multiplication to find the area C is 21 square units.

Choice B correctly identifies a figure with an area of 24 square units. Your child can demonstrate that they understand area by counting the unit squares individually, multiplying the number of rows by the number of columns, or by using addition.

EXAMPLE 20

Renee's office has a total area of 45 square feet and is shaped like a rectangle. Which figure could represent Renee's office?

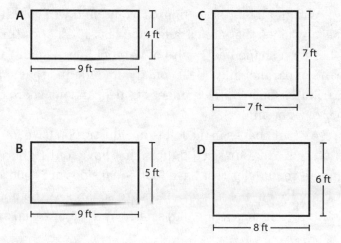

Discussion

Answer B contains the only two width × length measurements that will produce an area of 45 square feet. Students can find the area of each rectangle using multiplication, or by partitioning the lengths and widths by drawing additional lines to create same-sized squares that they can add in groups, or count individual squares. Students fluent with multiplication facts that also understand they are looking for a product equaling 45 may realize that none of the dimensions for either C or D include factors of 45.

Geometry

Third graders add the attributes of shapes to their geometry vocabulary and use these attributes to sort, identify, compare and contrast, and categorize shapes. A high-level classification of shapes are quadrilaterals. Whereas *polygon* refers to many-sided figures, *quadrilaterals* are classified as shapes with exactly four sides.

FACT

All four-sided closed figures are quadrilaterals. A *rhombus* is a quadrilateral that has four equal sides. A *square* is a special kind of rhombus where all four angles are right angles. Therefore, a square is a rhombus, but not all rhombuses are squares.

Within all four-sided figures, your child will identify whether the shape has any sides that are parallel, and classify shapes further by determining whether a shape has 2 pairs of parallel lines, 1 pair of parallel lines, or no pairs of parallel lines. He'll come to understand that shapes will share some attributes (such as the number of sides) and not share others (like the number of right angles).

All four shapes in the following illustration have four sides and are quadrilaterals. All four-sided figures that have four right angles are rectangles and all rectangles that have four equal sides are squares, so all squares are rectangles but not all rectangles are squares. A rhombus (the second figure from the left) has four equal sides, but not four equal angles.

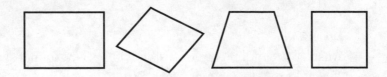

Your child will often be taught to subdivide a shape into smaller shapes to make problem solving more manageable. Because multiplication and division operations are introduced in third grade, students are limited to finding the areas of rectangles and squares; calculations for finding the area of other shapes is left to later grades.

EXAMPLE 21

Circle all of the shapes that are parallelograms.

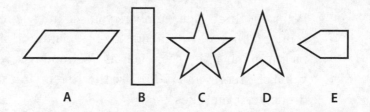

A   B   C   D   E

Discussion

A parallelogram must have four sides (and are quadrilaterals). So shapes C and E are *not* parallelograms. Parallelograms have two pairs of opposite parallel sides. Shape D does not have any parallel sides. Both A and B have four sides, and each have two pairs of parallel opposite sides so they are both parallelograms.

Helping Your Child Succeed

Third graders need encouragement. Multiplication, division, and fractions are all topics that students can struggle with for a long time. Providing students with positive reinforcement, encouraging them not to give up, or suggesting they approach problems in another way are all very important. Though math should be all about process and reasoning and not memorization, helping your child become fluent with the single-digit multiplication tables will benefit her for the rest of her life.

The area models presented in third grade are intended to help students develop procedural fluency with multiplication, so that multiplying becomes an easily accessible tool for solving problems. If students don't have procedural fluency with multiplication and division with numbers up to 100 by the time they reach middle school, they risk not being on track for being college- and career-ready. You can also remind your child that mastery is *not* expected the first time that a concept is introduced.

Around the House

There are lots of things you can do as part of your daily routine to reinforce the math concepts your child is learning in school. Make it fun for both you and your child. Here are some ideas for incorporating math into every-day events:

- Point out three-digit numbers you see anywhere—house numbers, route and highway numbers, license plates, or prices. For example, pick a number you see, such as 782, and ask your child what the value of 7 represents (700). Use this to reinforce place value for hundreds, tens, and ones.
- While you're waiting for dinner at a restaurant or sitting in the dentist's waiting room, write out a three-digit number, such as 544, and ask your child to round it to the nearest 100, or nearest 10. Use this to reinforce place value for hundreds, tens, and ones.
- Play "counting up and down" games using the numbers you see, and ask your child how much would have to be added or subtracted to get to the nearest 100.
- Play online and offline games that use math but are not specifically centered around math. Any game or puzzle that exercises logic and reasoning builds math skills. SET is an outstanding card game that builds discrete math skills and doesn't use numbers or arithmetic operations. Information and an effective tutorial can be found at *Setgame.com*.

Asking Questions

Learning to tell time can be fun, and you can work on it casually throughout the day. "How long do you think it will take to get to school?" "We are leaving at 5:00. How many minutes until we leave?" Other measurements can also find their way into your day: "How many tenths of a mile is it to school?" You can introduce kids to measuring using measuring spoons, measuring cups, and tape measures.

Talking about math, having students show all of their work, and answering questions about their thought process all help to deepen math understanding. It is often easier to solve a problem than to explain to someone how to solve the problem. When students need to explain how to solve a problem they need to organize their thoughts, select and use vocabulary, and demonstrate their underlying understanding. You can help your child succeed by asking him to explain his answer in more detail. Other question-asking techniques include:

- Repeat what your child says to give him a chance to listen to his own answer. This may trigger him to add additional information, reveal an error, or address a misconception. You can begin this process with "So what you are saying is . . . ."
- Propose a solution path; ask your child, "Do you think this would work? Why (or why not)?"

Exercises

1. What unknown number makes this equation true?

   $505 - 256 = \underline{}$

2. What unknown number makes this equation true?

   $32 \div 8 = \underline{}$

3. What unknown number makes this equation true?

   $9 \times \underline{} = 54$

4. Jill and Gabby are counting their stuffed animals. Gabby has 121. Magic the dragon is her favorite. Jill has 78 stuffed animals. Her favorite is Princess, a purple teddy bear. How many more stuffed animals does Gabby have than Jill?

5. What is 449 rounded to the nearest 100?

6. What is 823 rounded to the nearest 10?

7. Niko is collecting soda cans to recycle. He can redeem each can to get 5 cents. When he has 400 cans, he can redeem them to pay for a new video game. If Niko has 276 cans now, how many more must he collect before he has enough for a new video game?

8. Which fraction is equivalent to $\frac{2}{4}$?

   A. $\frac{1}{2}$ C. $\frac{1}{4}$

   B. $\frac{1}{3}$ D. $\frac{1}{6}$

9. Steve has an office that measures 48 square feet and is shaped like a rectangle. Draw and label a diagram showing the possible measurements of Steve's office.

10. Tommy asked kids in his class who their favorite basketball team was. Use the bar graph to answer the question.

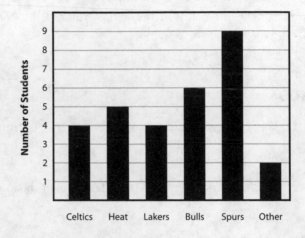

Students Favorite Basketball Team

How many more students said their favorite team was the Spurs than students who said the Heat?

11. Jill's pencil has a mass of 25 grams. Her pen has a mass that is 11 grams more than her pencil. What is the mass of her pen, in grams? Jill said that 3 of her pencils would have exactly the same mass as two of her pens. Do you agree with Jill? Why or why not?

12. A city park in Greenville is in the shape of a rectangle. The park is 140 feet long and 51 feet wide.

51 feet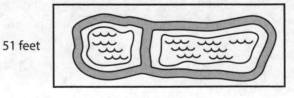

140 feet

Find the perimeter of the park in feet.

13. What time is shown on the clock?

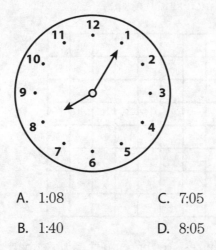

A. 1:08   C. 7:05

B. 1:40   D. 8:05

14. Circle all of the shapes that are quadrilaterals.

15. Circle each shape that is a rhombus.

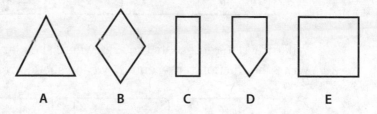

16. Bobby has a 50-pound bag of sand. He filled buckets of sand and used a scale to make sure that he had the same amount in each bucket. If each bucket contained 10 pounds of sand, how many buckets did Bobby fill?

17. Samantha takes 6 friends rollerblading to celebrate her birthday. Each person wears 2 rollerblades. Each rollerblade has 4 wheels. How many wheels do Samantha and her friends have all together?

18. Juan's father made him a chessboard out of wood. Each square on the board has a 2-inch side length.

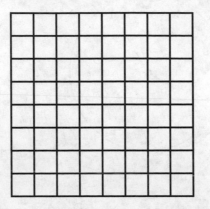

A. How many squares are there on the chessboard?

B. What is the area of each square?

C. What is the area of the chessboard?

19. Mrs. Simons planted a garden and partitioned it into 4 equal areas.

What fraction represents the area where she planted corn?

20. Write the fraction represented by each shaded region.

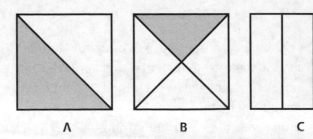

A   B   C

CHAPTER 8

Grade 4

Students in third grade perform multiplication using numbers 1–10, and perform division of numbers up to 100, but fourth graders will multiply and divide larger multi-digit numbers. Your fourth grader will also build upon his understanding of fractions by finding equivalent fractions, performing addition and subtraction with fractions having the same denominators, and multiplying fractions by whole numbers. Additionally your child will learn and apply geometry vocabulary such as *parallel*, *perpendicular*, and *symmetry*, and how to measure angles. It's going to be a busy year; let's get started!

What Your Fourth Grader Is Expected to Learn

Fourth grade is a year with a lot of new material to learn. In fourth grade your child will:

- Perform multi-digit operations and increase her understanding of place value up to 1,000,000.
- Extend her understanding of whole numbers and the properties of operations of whole numbers.
- Use decimal numbers to represent the fractions $\frac{1}{10}$ and $\frac{1}{100}$.
- Add and subtract fractions, and multiply fractions by whole numbers.
- Make use of visual fraction models and equations to represent and solve word problems. (Word problems in fourth grade involve a one-step or two-step process, and may require using more than one arithmetic operation.)

Common Core State Standards: Grade 4

There are five general categories of math skills your child will be learning in fourth grade, and certain abilities that fall under each of those categories:

OPERATIONS AND ALGEBRAIC THINKING
- Use the four operations with whole numbers to solve problems.
- Gain familiarity with factors and multiples.
- Generate and analyze patterns.

NUMBER AND OPERATIONS IN BASE TEN
- Generalize place value understanding for multi-digit whole numbers.
- Use place value understanding and properties of operations to perform multi-digit arithmetic.

NUMBER AND OPERATIONS—FRACTIONS
- Extend understanding of fraction equivalence and ordering.
- Build fractions from unit fractions by applying and extending previous understandings of operations on whole numbers.

- Understand decimal notation for fractions, and compare decimal fractions.

MEASUREMENT AND DATA
- Solve problems involving measurement and conversion of measurements from a larger unit to a smaller unit.
- Represent and interpret data
- Geometric measurement: Understand concepts of angle and measure angles.

GEOMETRY
- Draw and identify lines and angles, and classify shapes by properties of their lines and angles.

For the full list of Common Core State Standards for Mathematics, see Appendix A.

What Your Child Should Know Before Fourth Grade

Before entering fourth grade, there are several skills your child should already possess. For example, your child should know the times tables from memory; that is, she should know how to multiply two single-digit numbers. She should also know how to add, subtract, multiply, and divide numbers up to 100 and have some ability to solve one-step and two-step equations. She may be comfortable adding and subtracting numbers greater than 100 and up to 1,000.

FACT

If your child has some understanding of unit fractions before entering fourth grade, such as $\frac{1}{3}$ and $\frac{1}{2}$, then she will have less difficulty adding unit fractions such as $\frac{1}{8} + \frac{1}{8} + \frac{1}{8} = \frac{3}{8}$, or decomposing fractions such as $\frac{2}{4}$ to $\frac{1}{4} + \frac{1}{4}$.

At this point your child can probably round numbers to the nearest 10, or nearest 100. She should have some sense of place value, and understand that the 4 in 439 represents 4 hundreds, or 400, and that the 3 represents 3 tens, or 30.

In terms of geometry, your child should also know basic two-dimensional and three-dimensional shapes and be able to describe their attributes. She should have a general idea that *perimeter* is the distance all the way around an object, and she should understand *area* as the number of unit shapes that a flat shape takes up.

Operations and Algebraic Thinking

Your child has been solving problems that involve unknown values since pre-K; unknown values have been represented with question marks, blanks, or squares. In fourth grade, students begin to use *variables*, a single letter that represents unknown values. The variable could represent an unknown value in an equation, or it could represent an unknown side length of a geometric figure.

Problem Solving with Addition, Subtraction, Multiplication and Division

Your child learns to solve multistep word problems by turning the verbal statements into equations, often with variables and often requiring more than one of the basic arithmetic operations. He continues to see multiplication as having a number of groups, where each group contains some number of objects. Your child learns to interpret word problems to identify the groups, and the objects within those groups, or which number represents the quantity and which number tells how many times.

EXAMPLE 1

Tony had 43 stamps. Bonnie has 3 times as many stamps as Tony. Which of the following equations can be used to find b, the number of stamps that Bonnie has?

A. $43 = b \times 3$ **C.** $43 + 3 = b$

B. $b = 43 \times 3$ **D.** $b = 43 \div 3$

Discussion

Multiple-choice questions can be an effective means for having your child think about what a correct and incorrect answer should look like. The correct answer is $b = 43 \times 3$. Using a more literal translation of the word problem, your child may look for $43 \times 3 = b$; problems such as this one give your child practice in looking on either side of the equal sign.

In this example the group size is given (43) and the number of groups is given (3). What is unknown is the product. The types of multiplication problems that your child will encounter can fall into one of the following formats, and may just be one step in solving a multistep problem.

- **The product is unknown:** Tony has 43 stamps. Bonnie has 3 times as many. How many stamps does Bonnie have? $43 \times 3 = b$
- **The group size is unknown:** Bonnie has 129 stamps; that is 3 times more than Tony. How many stamps does Tony have? $129 = t \times 3$, or $129 \div 3 = t$
- **The number of groups is unknown:** Bonnie has 129 stamps. Tony has 43 stamps. How many times more stamps does Bonnie have than Tony? $129 = x \times 43$

EXAMPLE 2

Mrs. Kelly the English teacher bought 9 packages of blue pens and 7 packages of pencils. Each package of pens contained 12, and each package of pencils contained 10. How many more pens did Mrs. Kelly buy than pencils?

Discussion

Students can choose to model the problem representing the packages as groups of pens or pencils. They will need to first answer these two questions: How many pens did Mrs. Kelly purchase? How many pencils did Mrs. Kelly purchase? Then they can compare the difference.

She bought 9×12 pens $= 108$ pens.

She bought 7×10 pencils $= 70$ pencils.

She bought $108 - 70 = 38$ more pens than pencils.

Students will see some information in the problem that is not important, such as the name of the teacher, the subject that she teaches, and the color of the pens. A common error is to calculate the total number of pens and the total number of pencils that where purchased and forgetting to really answer the question, which is finding the difference.

ESSENTIAL

An estimation problem might ask, "About how many more pens did Mrs. Kelly buy than pencils?" Using mental math, your child could round the number of packages of pens purchased *up* to 10, and calculate that she bought about 120 pens. The number of pencils she bought can be quickly found by multiplying 7 and 10 to get 70, and a quick estimate of how many more pens than pencils she bought is 120 − 70 = 50. By rounding up, making the number of pens higher than actually purchased, your child should expect that the actual difference between the number of pens bought and the number of pencils will be *less* than 50.

Working with Factors and Multiples

In fourth grade, your child differentiates between prime numbers and composite numbers, and learns about factors and multiples. In third grade, he learned multiplication and maybe skip-counting. Fourth-grade students become familiar with the vocabulary terms *multiples* and *factors* of whole numbers through 100. In future grades, he will realize that a number is either a prime number, or can be represented as a product of prime numbers.

FACT

The number 1 is neither prime nor composite. A *prime number* is greater than 1, and can be divided, without a remainder, by only itself and 1. For example, 17 can be divided only by 17 and by 1. No prime number greater than 5 ends in a 5, because any number greater than 5 that ends in a 5 can be divided by 5. Two is the only *even* prime number; all other even numbers can be divided by 2. A *composite number* is any number greater than 1 that is not prime, and therefore has more than two factors.

Your child will learn a variety of strategies to find the factors of numbers from 1 to 100, including drawing factor trees, using skip-counting, making tables, drawing diagrams, and other methods for identifying factor pairs. Two numbers are factors of a third number if their product is the third number. For example, 4 and 6 are both factors of 24 because their product is 24. All of the factors of both 4 and 6 are also factors of 24.

Your child can use skip-counting techniques using known factors to test other numbers to see if they are also factors of 24. Since 6 is a factor of 24, and 6 has factors of 2 and 3, numbers to test to see if they are also factors include 2, 4, 6, 8, 10, 12, 3, 6, 9, 12, and 15. Your child will eventually develop a recognition of where to stop testing. In this last case, 15 would have to be multiplied by a number at least as high as 2, and 15×2 is 30, which is greater than 24, so all numbers greater than 2 will also provide a product greater than 24.

EXAMPLE 3

List all of the prime numbers from 50 to 75.

Discussion

If your child realizes that no even prime numbers exist between 50 and 75, then she could start this problem by listing all of the odd numbers from 51 to 75:

51, 53, 55, 57, 59, 61, 63, 65, 67, 69, 71, 73, 75

She could then use any other divisibility facts she knows, such as if a number ends with a 5 or 0 then it is divisible by 5, so 55, 65, and 75 can be eliminated. Next, she might remove multiples of known prime numbers (or multiples of composite numbers). Deleting the multiples of 3 (51, 57, 63, and 69) would leave 53, 59, 61, 67, 71, and 73.

Your child could then test each remaining number to see if they are divisible by 7 (or check multiples of 7), and see that none remain. As she tests 7, she may see that 7×11 is beyond the highest remaining number, and that if any number in this set were a composite number, then one of those numbers must be less than 11. With practice, she will realize when she has tested for all possible numbers.

Generating and Analyzing Patterns

In fourth grade, your child analyzes sequences of numbers or shapes, and is able to extend them. She could analyze addition or multiplication tables and find very interesting patterns. She may also use tables and follow rules to generate additional (or missing) numbers from a sequence, and will be given a sequence and asked to describe the behavior of the rule.

EXAMPLE 4

What is the next number in this pattern?

3   5   9   17   33   65

Answer
129

Discussion

Your child may "see" the pattern, or apply the strategies she has learned to identify the pattern. She may first examine the difference between the numbers or see if somehow they are related by multiples. The first number is 3, the next number is 2 more; the second number is 5, the next number is 4 more. The third number is 9, the next number is 8 more. She could take this horizontal list of numbers and create a vertical list, and she might also number the numerals in the list to see the patterns.

For this example, the next number is found by doubling the previous number and subtracting one. Your child may summarize this as "I had 3, I added 2; I had 5, I added 4; I had 9, I added 8; . . . I had 33, I added 32; I had 65, I added 64."

Number and Operations in Base Ten

Fourth graders perform operations on multi-digit numbers that can reach as high as 1,000,000, so they must gain an understanding of place value using more digits. Your child should be able to compute using more place values, and be able to compute using more *operands*. To be able to understand the number 55,678 he must be able to add the five numbers $50,000 + 5,000 + 600 + 70 + 8$.

ESSENTIAL

An *operand* is the object of the operation, the quantity on which the operation is being performed. For example, in the equation $2+5=7$, the operation being performed is addition. The 2 and the 5 are the operands because they are the objects that are being added; they are the objects of the addition operation.

Understanding Whole Numbers

In fourth grade your child gains an understanding of larger, multi-digit whole numbers using place value, and how each digit of a multi-digit number relates to the number on the left and the right. For the first time, your child will use the comma to help him separate the digits of large numbers to make them easier to read and to write. He represents multi-digit numbers using number names, numerals, and expanded form, and expands his use of estimation using his understanding of place value.

When multiplying by 10 your child may add a 0 to the end of a number, such as $12 \times 10 = 120$, and when he divides a number such as 870 by 10, he may simply remove the zero to get 87.

EXAMPLE 5

Complete the following table.

Example:	3 thousands =	3000 ones =	300 tens =	30 hundreds
	8 hundreds =	_____ ones =	_____ tens	
	40 tens =	_____ ones =	_____ hundreds	
	600 ones =	_____ tens =	_____ hundreds	

Discussion

When your child converts from ones to tens to hundreds, he sees place value relationships between multiplying and dividing by ten. He may need to see the equal sign as meaning *the same as*, and can use place value relationships when performing operations.

Students in fourth grade work with numbers up to 1,000,000, read and write number names of large numbers, and represent them with numerals and in expanded form. For example, $84,323 = 80,000 + 4,000 + 300 + 20 + 3$. Fourth graders also learn how to read numbers containing commas, understanding that 325,800 is read as "three hundred twenty five thousand eight hundred."

EXAMPLE 6

Compare the following numbers. Fill in the blanks using >, =, and < to show the relationship.

A. 339 _____ 300 + 40 + 8

B. Six thousand one hundred seventy one _____ 6,871

C. 10,000 + 2,000 + 30 + 6 _____ 10,000 + 400 + 60 + 7

D. 12,336 _____ 10,467

E. 876,471 _____ 471,867

Discussion

Your child needs to relate verbal models (the number names) to the numeric value of numbers and their expanded form. When he is comparing multi-digit numbers, he can use his understanding of place value to make appropriate choices between <, =, or > without having to convert to a common representation. He may have the most difficulty with C if he doesn't see that 12,036 has 0 represented in the hundreds position and 10,467 has 0 represented in the thousands position. He may also have difficulty with E if he doesn't understand the use of the comma.

Rounding

Being able to round numbers to any place—ten, hundreds, or thousands—is a great tool for using estimations. Your child can use estimations to check the reasonableness of his answers or even to monitor his own progress while solving multistep problems. He can use rounding to help explain his answers, and also when problems are asking for approximate answers.

EXAMPLE 7

A football stadium with 65,000 people has 10 exits. Tommy counts that 160 people left using 1 exit in 1 minute. About how long would it take for everyone to leave the stadium?

Discussion

Using place value, Tommy could estimate that 1,600 people would leave the stadium using all exits in one minute. Then 16,000 people would leave the stadium in 10 minutes; 32,000 would leave the stadium in 20 minutes; and 64,000 people would leave in 40 minutes. That would leave 1,000 people in the stadium, but according to Tommy's first estimate, it would take less than a minute for that many people to leave, so 40 minutes is a good estimate.

Performing Multi-Digit Arithmetic

Your child can leverage his understanding of place value to simplify the calculations performed with multi-digit numbers. In fourth grade, students use the standard algorithms for addition and subtraction with multi-digit numbers. Using strategies such as area models to represent multiplication and division will prepare your child for learning the standard algorithms for multiplication and division in later grades.

An *area model* is a rectangle divided into rows and columns to demonstrate multiplication and division.

EXAMPLE 8

There are 52 cards in a standard deck of playing cards. Use an area model to find out how many cards there are in one dozen decks of playing cards. Explain your answer.

Discussion

There are 12 in one dozen, so the number of cards can be found using 12×52. An area model uses the fact that $12 = 10 + 2$ and $52 = 50 + 2$, so the area model is used to find the sum of four products.

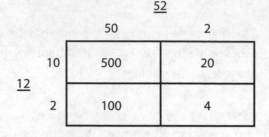

The number of cards in the 12 decks is $500 + 100 + 20 + 4 = 624$.

Breaking down 52 and 12 allows your child to work with large numbers, but at the same time uses numbers that he can manage using mental math. He'll gain proficiency in multiplying single-digit numbers, then multiply single-digit numbers by 10, and then multi-digit numbers by 10. The area models asks him to find the sum of products that he can manage. He can apply multiple strategies for finding the sum of the four products; in the example the hundreds were added together, then the tens, and then the ones.

Connecting Place Value and the Standard Algorithms

You probably remember using terms such as *carrying* and *borrowing* when adding and subtracting. Students today use words like *grouping* and *ungrouping*. The standards refer to this as *composing* and *decomposing* numbers. Your child will use an understanding of place value and the properties of operations to transition toward using the standard algorithms for addition and subtraction. He will also deepen his understanding of the relationships between addition and subtraction.

EXAMPLE 9

Bryan scored 4,548 playing a video game, and Luke scored 3,885. How many points did they score all together?

Discussion

Together they scored $4548 + 3885$. Using place value, this can be expressed as:

$$4000 + 500 + 40 + 8 + 3000 + 800 + 80 + 5$$

Using the properties of operations, in particular the commutative property, this can be rearranged to group the thousands, hundreds, tens, and ones, and becomes:

$$4000 + 3000 + 500 + 800 + 40 + 80 + 8 + 5$$

This simplifies to:

$$7000 + 1300 + 120 + 13$$

From here students can add these four numbers however they want, maybe $8300 + 133 = 8433$.

Using a vertical alignment of adding these in expanded form looks more like the standard algorithm:

$$\begin{array}{l} 4000 + 500 + 40 + 8 \\ + 3000 + 800 + 80 + 5 \\ \hline \end{array}$$

Or

$$\begin{array}{l} 4 \text{ thousands} + 5 \text{ hundreds} + 4 \text{ tens} + 8 \text{ ones} \\ + 3 \text{ thousands} + 8 \text{ hundreds} + 8 \text{ tens} + 5 \text{ ones} \\ \hline 7 \text{ thousands} + 13 \text{ hundreds} + 12 \text{ tens} + 13 \text{ ones} \end{array}$$

13 ones can rewritten as 1 ten plus three ones, so now there are 13 tens and three ones

$$7 \text{ thousands} + 13 \text{ hundreds} + 13 \text{ tens} + 3 \text{ ones}$$

13 tens can be rewritten as 1 hundred plus 3 tens, so now there are 14 hundreds plus 3 tens

$$7 \text{ thousands} + 14 \text{ hundreds} + 3 \text{ tens} + 3 \text{ ones}$$

14 hundreds can be rewritten as 1 thousand plus 4 hundreds, so now there are 8 thousands, plus 4 hundreds plus 3 tens plus 3 ones

$$8 \text{ thousands} + 4 \text{ hundreds} + 3 \text{ tens} + 3 \text{ ones}$$
$$8433$$

EXAMPLE 10

Bryan scored 4,548 playing a video game, and Luke scored 3,885. How many more points did Bryan score than Luke?

Discussion

One way to find how many more points Bryan scored than Luke is to ask, "How many more points are needed for them to score the same?" Some students' first instinct is to change all subtraction problems to addition, and then use a strategy such as the count up method to answer the question. In this case they would answer the question $3885 + ? = 4548$.

They would add 15 to get to 3900, then add 100 to get to 4000, then add 548 to get to 4548. Then they would add $15 + 100 + 548$ to get 663, and answer that Bryan scored 663 points more than Luke.

Setting this up to subtract two numbers in expanded form looks like:

$$\begin{array}{r} 4000 + 500 + 40 + 8 \\ -\ 3000 + 800 - 80 + 5 \\ \hline \end{array}$$

And working from right to left, subtracting one place at a time, $8 - 5 = 3$, so there are three ones. It is not possible to subtract 8 tens from 4 tens, so one of the 5 hundreds must be regrouped. After regrouping, this becomes 4 hundreds and 14 tens, and 14 tens $-$ 8 tens is 6 tens:

$$\begin{array}{r} 4000 + 400 + 140 + 8 \\ -\ 3000 + 800 - 80 + 5 \\ \hline 60 + 3 \end{array}$$

It is not possible to take 8 hundreds away from 4 hundreds, so one of the 4 thousands must be regrouped to become 3 thousands and 14 hundreds. Then it is possible to take away 8 hundreds from 14 hundreds, which is 6 hundreds.

$$\begin{array}{r} 3000 + 1400 + 140 + 8 \\ -\ 3000 + 800 - 80 + 5 \\ \hline 600 + 60 + 3 \end{array}$$

The final answer is 663.

EXAMPLE 11

Show how David could use an area model to solve the following problem.

Workers want to pour a $\frac{1}{4}$ mile sidewalk in one 8-hour day. That is 1320 feet of concrete. How many feet of concrete do they need to pour in each hour?

David knew that the inside area of his model needed be represent the 1320 feet, and the left side of his model would represent the 8 hours. He wasn't sure what should be on the top of the model so he added ? + ?	? + ? 8 | 1320 |
He noticed that they needed to pour at least 100 feet per hour because 1320 > 8×100, so he added an area for 100 feet each hour.	100 ? 8 | 800 | | 1320 - 800 520
$1320 - 800 = 520$. David saw that 520 is more than half of 800, so they would need to pour at least 50 more feet per hour.	100 50 ? 8 | 800 | 400 | | 1320 520 - 800 - 400 520 120
$520 - 400 = 120$. David saw that 120 is more than 80, so they would need to pour at least 10 more feet each hour.	100 50 10 ? 8 | 800 | 400 | 80 | | 1320 520 120 - 800 - 400 - 80 520 120 40
$120 - 80 = 40$. David saw that 40 is 8×5, so they would need to pour exactly 5 more feet each hour.	100 50 10 5 8 | 800 | 400 | 80 | 40 | 1320 520 120 40 - 800 - 400 - 80 - 40 520 120 40 0

David's final total was $100 + 50 + 10 + 5 = 165$ feet each hour.

Remainders

In fourth grade, your child will encounter division problems that have remainders. But there is no need to panic. One thing your child will have to understand is that having remainders is not dependent on the strategy that was used.

EXAMPLE 12

Solve $412 \div 3$.

Discussion

If there were 412 items to be placed in 3 buckets, first put 100 in each bucket. That leaves 112. If you put 30 more in each bucket, that leaves $112 - 90 = 22$. Next put 7 in each bucket, which leaves 1 left over because $7 \times 3 = 21$. The total number of objects in each bucket is $100 + 30 + 7 = 137$. There is one left over, so the remainder is 1.

FACT

The second Standard for Mathematical Practice is "Reason abstractly and quantitatively," and within this practice are the concepts *contextualize* and *decontextualize*. The previous problem is just a calculation; putting the problem into a context helps your child to understand division as partitioning an amount into equal groups.

Number and Operations—Fractions

In fourth grade your child extends her understanding of whole numbers and operations to fractions. Your child is learning to add and subtract fractions with the same denominator and to multiply fractions by whole numbers. That is what she is asked to *do*, but what she is asked to *understand* is a bit more complex. Here is a summary of the vocabulary used by the Common Core writers to express how your child will extend her understanding of whole numbers and operations to fractions.

Vocabulary	Example(s)	Notes
Numerator / Denominator	$\dfrac{3}{4}$	The top number, the *numerator*, represents the number of pieces being considered, out of the number of parts of the whole, called the *denominator*. This fraction describes 3 parts of a whole that has been partitioned into 4 equal parts.
Unit fraction	$\dfrac{1}{2}, \dfrac{1}{3}, \dfrac{1}{8}$	Fractions that have 1 as a numerator are called *unit fractions*. They represent one part of a whole.
Referring to the same whole	$\dfrac{1}{2}$ $\dfrac{1}{4}$	Students will not be asked to operate on fractions that do not refer to the same whole, such as $\dfrac{1}{4}$ of and $\dfrac{1}{2}$ of
Having like denominators	$\dfrac{1}{4} + \dfrac{3}{4}$	Before operating on fractions that do not have the same denominator, such as $\dfrac{1}{2} + \dfrac{1}{4}$, your child may be asked to first find (and use) *equivalent fractions* such as $\dfrac{2}{4} + \dfrac{1}{4}$.
Decomposing a fraction	$\dfrac{3}{8} = \dfrac{1}{8} + \dfrac{1}{8} + \dfrac{1}{8}$	Your child might not use the phrase "decomposing a fraction," but he will experience it using a visual fraction model or an equation.
Using properties of operations	$\dfrac{2}{3} + \dfrac{1}{3} = \dfrac{1}{3} + \dfrac{2}{3}$	The commutative property of addition states that the order of the numbers being added (the addends) can be rearranged.
Using visual fraction models and equations	$\dfrac{2}{4} + \dfrac{1}{4} = \dfrac{3}{4}$	Fraction models and equations can be used separately or together.

Keep in mind that your child is not learning all this information about fractions in one day. Working with fractions supports your child's understanding of operations and whole numbers.

Equivalent Fractions

Fourth-grade students do not have experience with multiplying fractions, so visual fraction models are used to find equivalent fractions.

Your child may represent equivalent fractions using models such as this to show that the area covered by $\dfrac{1}{2}$ is the same as the area covered by $\dfrac{2}{4}$.

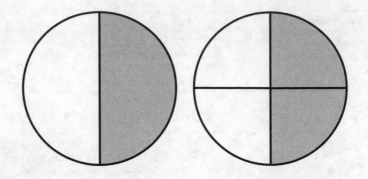

Alternatively, he could show that two fractions represent the same point on a number line.

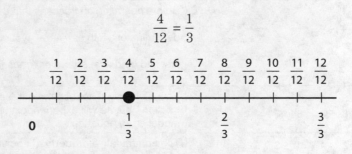

Your child's experience with modeling fractions should help him to see patterns for creating equivalent fractions.

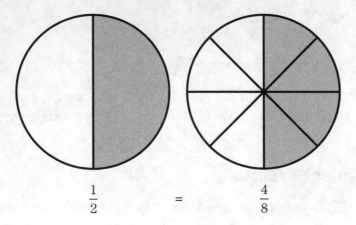

There are four times as many shaded regions, and there are four times as many parts to the whole.

$$\frac{4 \times 1}{4 \times 2} = \frac{4}{8}$$

When your child compares fractions, he may not need to find equivalent fractions; instead, he may use different strategies. For example, if your child is comparing $\frac{2}{3}$ and $\frac{3}{8}$, he may use a third number, such as $\frac{1}{2}$, to use as a baseline. He may realize that $\frac{2}{3}$ is greater than $\frac{1}{2}$ while $\frac{3}{8}$ is less than $\frac{1}{2}$. Mentally, he puts all three numbers in order: $\frac{3}{8}$, $\frac{1}{2}$, $\frac{2}{3}$ to find that $\frac{3}{8} > \frac{2}{3}$.

Another strategy for comparing fractions such as $\frac{2}{3}$ and $\frac{3}{8}$ without finding equivalent fractions is to plot each fraction on the number line. Your child learns to read number lines with whole numbers in order to recognize that the least of two numbers is positioned closer to zero on a number line (or on the left), and the greater of two numbers is the farthest from zero on a number line (or the number to the right). The same relationship is true of fractions.

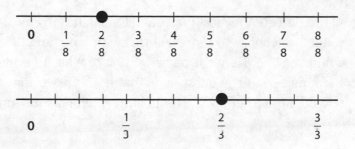

EXAMPLE 13

Declan and Jordan each bought a medium pizza. Declan cut his pizza into eighths and ate 3 slices. Jordan cut his pizza into fifths, and ate 3 slices. Draw a diagram showing how much pizza each boy has left. Who ate the most pizza? Explain your answer.

Discussion

Declan cut his pizza into 8 slices. He ate 3 slices, and has 5 left.

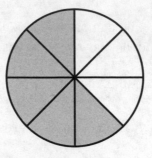

Jordan cut his pizza into 5 slices. Jordan ate 3, and has 2 left.

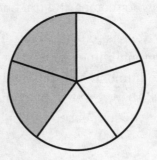

Jordan ate the most pizza, and Declan has the most pizza left. Your child could also choose to represent this with a rectangular array/tape ribbon or a number line. He must pay attention to what the diagram is supposed to show, and what the question is that he is answering, and how the two are related. For either pizza, the amount eaten plus the amount left must equal one pizza.

Building Fractions from Unit Fractions

Your child will use his prior knowledge about whole numbers and properties of operations to build fractions from unit fractions. By understanding unit fractions, your child can connect multiplying a fraction by a whole number to the concept of repeated fraction addition.

Your child will use a combination of visual fraction models and equations to see addition as joining together parts that refer to the same whole.

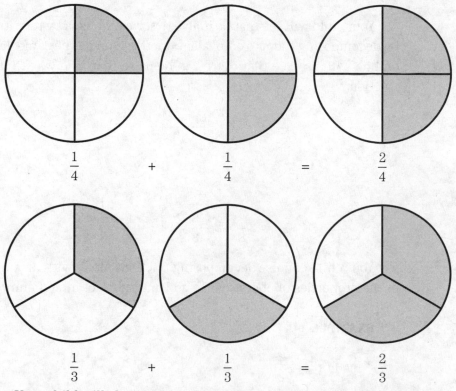

Your child will also use a combination of visual fraction models and equations to see subtraction as separating parts that refer to the same whole.

Mark had $\frac{2}{3}$ of a pie left, then he ate $\frac{1}{3}$ of the pie for dessert. How much of the pie does he have left?

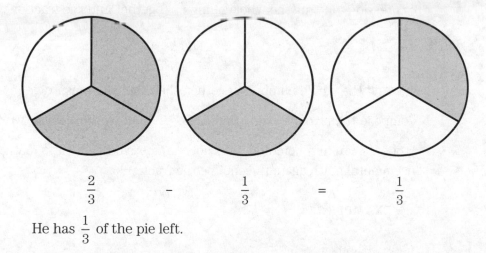

He has $\frac{1}{3}$ of the pie left.

Your child will see that he is able to solve problems by rewriting fractions by decomposing them to equivalent fractions (using the same denominator), and he will also see that there can be more than one way to decompose a fraction.

$$\frac{3}{8} = \frac{1}{8} + \frac{1}{8} + \frac{1}{8}$$

$$\frac{3}{8} = \frac{1}{8} + \frac{2}{8}$$

$$2\frac{3}{8} = 2 + \frac{3}{8} = 1 + 1 + \frac{1}{8} + \frac{1}{8} + \frac{1}{8}$$

Your child can use decomposing fractions along with his understanding of the properties of operations to add and subtract mixed numbers.

EXAMPLE 14

Add $3\frac{1}{8} + 2\frac{1}{8}$

Discussion

Decompose both mixed numbers and rewrite the expression as $3 + \frac{1}{8} + 2 + \frac{1}{8}$. Then use the commutative property of addition to reorganize the expression, putting the whole numbers together and the fractions together:

$$3 + 2 + \frac{1}{8} + \frac{1}{8}$$

Add the whole numbers together, then the fractions, to get $5 + \frac{2}{8}$. Then compose them to a mixed number: $5\frac{2}{8}$. When working with two fractions, students can first combine the whole numbers using mental math, and then use mental math again to combine the fractions.

EXAMPLE 15

Subtract: $3\frac{3}{8} - 2\frac{1}{8}$

Discussion

Your child could use a visual fraction model or an equation to solve this. A number line might be a better choice for this subtraction problem than a diagram with circles or boxes. One approach would simulate what would happen on a number line with an equation. Start with $3\frac{3}{8}$, subtract 2 to get $1\frac{3}{8}$, then subtract $\frac{1}{8}$ more to get $1\frac{2}{8}$.

Another approach is to decompose $3\frac{3}{8}$ to $3+\frac{3}{8}$ and $2\frac{1}{8}$ to $2+\frac{1}{8}$ and subtract the whole numbers: $3-2=1$. Then subtract the fractions: $\frac{3}{8}-\frac{1}{8}=\frac{2}{8}$. The final answer is the composition of the whole number 1 and the fraction $\frac{2}{8}=1\frac{2}{8}$.

EXAMPLE 16

This is a model of $\frac{11}{3}$. What is $\frac{11}{3}$ as a mixed number?

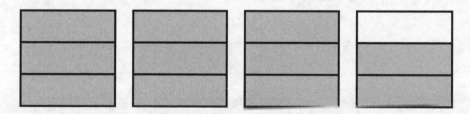

Discussion

The model shows the three wholes completely shaded in, and two of the three thirds are shaded in on the fourth whole. This represents $3\frac{2}{3}$.

Students are expected to be able to represent mixed numbers with a visual model, *and* to interpret a mixed number from a visual model.

Multiplying Fractions by Whole Numbers

Your child can extend his understanding of whole numbers and operations to fractions by seeing that multiplying a fraction by a whole number is the same as multiplying a whole number by a fraction; seeing a fraction being repeatedly added is a clear way to understand how to multiply fractions by whole numbers. First, look at the repeated addition of a unit fraction:

$$\frac{1}{8} + \frac{1}{8} + \frac{1}{8} + \frac{1}{8} + \frac{1}{8}$$

This represents $\frac{1}{8}$ five times, or $5 \times \frac{1}{8}$ which is $\frac{5}{8}$.

Now suppose that we have $\frac{5}{8}$ repeated a number of times, such as $\frac{5}{8} + \frac{5}{8} + \frac{5}{8}$. This represents $\frac{5}{8}$ repeated 3 times, or $3 \times \frac{5}{8}$ This could be decomposed into $\frac{1}{8} + \frac{1}{8} + \frac{1}{8} + \frac{1}{8} + \frac{1}{8}$ plus $\frac{1}{8} + \frac{1}{8} + \frac{1}{8} + \frac{1}{8} + \frac{1}{8}$ plus $\frac{1}{8} + \frac{1}{8} + \frac{1}{8} + \frac{1}{8} + \frac{1}{8} = \frac{15}{8}$.

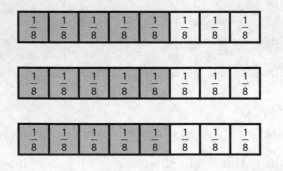

$$3 \times \frac{5}{8} = \frac{15}{8}$$

EXAMPLE 17

Mrs. Jenkins is planning a picnic for 9 people and wants to buy enough meat to make everyone a $\frac{1}{4}$ pound hamburger. How much hamburger should she buy?

Discussion

$$9 \times \frac{1}{4} = \frac{9}{4}$$

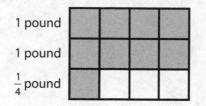

$$\frac{9}{4} = 2\frac{1}{4} \text{ pounds}$$

Each row of the rectangular array represents one pound of hamburger. Each row has been partitioned into four parts. Enough rows have been added so that 9 quarter-pound hamburgers could be represented. The shaded region represents two whole pounds of hamburger are needed and one additional quarter pound.

Representing Fractions with Decimals

In fourth grade your child begins to work with decimals, starting with representing the fraction $\frac{1}{10}$ with 0.1, and the fraction $\frac{1}{100}$ with 0.01. He is expected to convert fractions with a denominator of 10 to an equivalent fraction with a denominator of 100, such as $\frac{1}{10} = \frac{10}{100}$. This will help him to add two fractions that have denominators of 10 and 100. It is a first step toward both using decimal numbers and finding common denominators. Your child will also convert decimal numbers (up to two decimal places) to fractions, and compare decimal numbers by considering the meaning of a decimal number as a fraction.

A dime has the value of 10 cents, and there are 100 cents in one dollar. One dime represents $\frac{1}{10}$ of a dollars. Two dimes is $\frac{2}{10}$ of a dollar and 10 dimes is $\frac{10}{10}$ or one whole dollar. A penny has the value of one cent. There are 100 pennies in a dollar, so each cent is worth $\frac{1}{100}$ of a dollar. 10 cents represents $\frac{10}{100}$ of a dollar.

Students will notice that 10 cents can be represented by one dime, or 10 pennies. A decimal point is used to represent the separation of a whole number and a fraction. The digit to the right of the decimal point represents $\frac{1}{10}$.

EXAMPLE 18

Represent $\frac{1}{10}$, $\frac{2}{10}$, and $\frac{7}{10}$ using decimal numbers.

Discussion

$\frac{1}{10}$ as a decimal is 0.1. The 0 represents having 0 whole parts, and makes the number easier to read than .1. It is not incorrect to leave off the 0, but the numbers are more difficult to read. $\frac{2}{10} = 0.2$ and $\frac{7}{10} = 0.7$.

EXAMPLE 19

Represent 0.3, 0.4, and 0.9 as fractions.

Discussion

0.3 represents $\frac{3}{10}$, 0.4 represent $\frac{4}{10}$, and 0.9 as a fraction is $\frac{9}{10}$.

The place value to the left of the decimal points represents the number of ones, and the place value to the right of the decimal point represents the number of tenths ($\frac{1}{10}$). 0.3 is read as "three tenths," 0.4 is read as "four tenths," and 0.9 is read as "nine tenths."

Comparing Decimal Numbers

To compare decimal numbers, students may use number lines, compare fractions, or use visual models that represent decimal numbers. Here are three ways to represent 0.8.

As a fraction: $0.8 = \dfrac{8}{10}$.

On a number line:

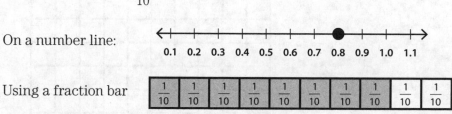

Using a fraction bar

EXAMPLE 20

Compare each pair of numbers using >, =, or <.

A. $\dfrac{3}{10}$ ____ 0.3

B. 0.9 ____ 0.4

C. $\dfrac{7}{10}$ ____ 0.2

Discussion

To compare $\dfrac{3}{10}$ to 0.3, either convert $\dfrac{3}{10}$ to a decimal or 0.3 to a fraction. Your child should see that $\dfrac{3}{10} = \dfrac{3}{10}$ or 0.3 = 0.3, so $\dfrac{3}{10} = 0.3$. To compare 0.9 to 0.4, he could plot both numbers on a number line, as shown in the previous example, or convert both to fractions to compare $\dfrac{9}{10}$ to $\dfrac{4}{10}$ and see that $\dfrac{9}{10} > \dfrac{4}{10}$. He can compare $\dfrac{7}{10}$ to 0.2 by comparing $\dfrac{7}{10} > \dfrac{2}{10}$, or by comparing 0.7 > 0.2.

A 10 × 10 grid can represent one whole, and each square represents $\dfrac{1}{100}$. Each row represents $\dfrac{1}{10}$; there are 10 rows. Similarly, each column represents $\frac{1}{10}$ and there are 10 columns. Because each column has 10 boxes, each column also represents $\frac{10}{100}$.

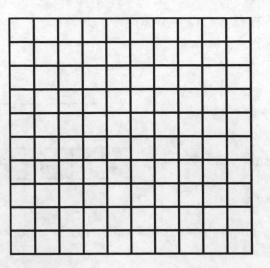

$\frac{3}{10}$ and $\frac{30}{100}$ are both represented by:

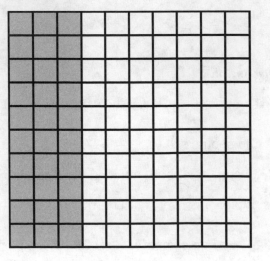

The second digit to the right of the decimal point represents $\frac{1}{100}$: 0.01.

$\frac{15}{100}$ is 0.15, and $\frac{75}{100}$ is 0.75. When comparing fractions such as $\frac{4}{10}$ and $\frac{56}{100}$, students have the option of comparing area models, changing $\frac{4}{10}$ to $\frac{40}{100}$ so that it can be compared to $\frac{56}{100}$, or using their positions on a number line.

Measurement and Data

In fourth grade, students convert units within the same system of measure from larger units to smaller units, such as from feet to inches. This prepares them in later grades for converting smaller units to larger units, as well as converting between systems of measures, such as from inches to centimeters. Fourth graders, becoming familiar with representing unknown values with variables, begin to use area and perimeter formulas. Fourth-grade students also learn how to measure angles using a protractor, and solve problems for unknown angle measures.

Converting Measurements

Your child will solve problems that convert length measures. Common measures that fourth graders will need to know include: 1 mile is 5,280 feet, 1 yard is 3 feet, and 1 foot is 12 inches. Metric system units that your child may use include: 1 kilometer is 1000 meters, 1 meter is 100 centimeters, and 1 centimeter is 10 millimeters.

EXAMPLE 21

Convert length measures from the world of sports.

A. A basket in basketball is 10 feet tall. How many inches is this?

B. A first down in football is 10 yards. How many feet is this?

C. The distance from second base to third base in baseball is 90 feet. How many inches is this?

D. The first hole at the Augusta National Golf Club where the Masters Tournament is played is 445 yards long. How many feet is this?

Discussion

There are 12 inches in one foot, so there are $10 \times 12 = 120$ inches in 10 feet.

There are 3 feet in one yard, so there are $10 \times 3 = 30$ feet in 10 yards.

There are 12 inches in one foot, so there are $90 \times 12 = 1{,}080$ inches in 90 feet.

There are 3 feet in the yard, so there are $445 \times 3 = 1{,}335$ feet in 445 yards.

These examples all converted whole units to whole units. Your child will likely encounter measurements that involve fractions or decimals, such as finding the number of inches in $4\frac{1}{2}$ feet.

ESSENTIAL

> The fourth-grade standards focus on converting from larger units to smaller units. When students create visual models, they should be able to calculate larger equivalent units from smaller units. Students will also solve unit-conversion problems that involve time, volume, mass, or money.

EXAMPLE 22

Judy bought a rectangular rug for her office that is 143 square feet. The length of the rug is 13 feet. Write an equation than can be used to find the width of the rug. Let w represent the width. What is the width of Judy's rug? Show your work.

Discussion

The area of a rectangle is found using $A = l \times w$. Here the area and the length are known, and the width is unknown. Even though the question asks for an equation, it is a good idea to also draw a diagram (a rectangle) and label the length 13 feet. If needed, draw 13 stripes in one direction, and additional stripes in the other and count by 13 until there are 143 blocks.

$$143 = 13 \times w$$

$$142 = 13 \times 11$$

The width of the rug is 11 feet.

ALERT

When using formulas, such as for the area of a rectangle, students must make sure that the units are consistent. For example, make sure that the length and the width are both specified in feet prior to multiplying to find the area (measured in square feet).

Represent and Interpret Data

Your child's skills for representing and interpreting data are developed each year. In fourth grade, students record data that can incorporate fractional units, such as $\frac{1}{2}$, $\frac{1}{4}$, and $\frac{1}{8}$. When students read data and perform calculations based on the data presented, they may also need to use fractional units.

EXAMPLE 23

Tammy helped measure the height of 12 dogs at the pet store and created a line plot of the data collected. What is the height difference between the tallest and shortest dog that Tammy measured?

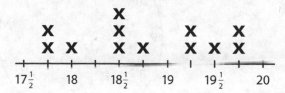

Dog Heights (inches)

Discussion

The height of the tallest dog Tammy measured was $19\frac{3}{4}$ inches. The height of the smallest dog was $17\frac{3}{4}$ inches. The difference is $19\frac{3}{4} - 17\frac{3}{4} = 2$ inches. Students in later grades will view this as the *range* of the data.

Geometric Measurement

An angle is formed when two rays meet at a common endpoint, such as the hands on a clock. Starting at 12:00, the angle formed by the hands of a clock increase 30 degrees each hour. At 3:00, the hands form a 90-degree (right) angle; at 4:00, the hands form a 120-degree angle. Angles are measured in reference to a full circle, which totals 360 degrees. Students in fourth grade learn to use a protractor to draw and to measure angles.

FACT

> A *ray* is a part of a line that begins at a particular point (called an endpoint) and extends in one direction.

EXAMPLE 24

Draw hands on the clock to show 5:00. Use a protractor to measure the angle formed by the two hands.

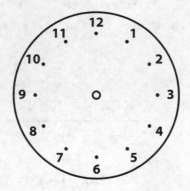

Discussion

At 5:00 the minute (big) hand is on the 12, and the hour (small) hand is on the 5. The angle formed is 150 degrees.

An angle that measures 90 degrees is called a *right angle*. A right triangle contains exactly one right angle. An angle that has a measurement of less than 90 degrees is called an *acute angle*. The sum of the measures of the three angles of all triangles is 180 degrees. In a right triangle, the sum of the measures of the two acute angles must be 90 degrees.

EXAMPLE 25

In triangle ABC, angle A is a right angle, with a measure of 90 degrees. Angle B is an acute angle that measures 30 degrees. What is the measure of angle C?

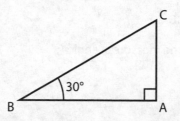

Discussion

The sum of the three angle measures of all triangles is 180 degrees.

$$180 = 90 + 30 + \ ?$$

Angle C must be 60 degrees. Two angles whose sum is 90 degrees are called *complementary angles*. Angle B and angle C are complementary angles.

When two angles are next to each other, sharing a side and an endpoint, they are called *adjacent angles*. When two angles are put side to side, the measure of the outside angle is calculated by adding the two angles. For example:

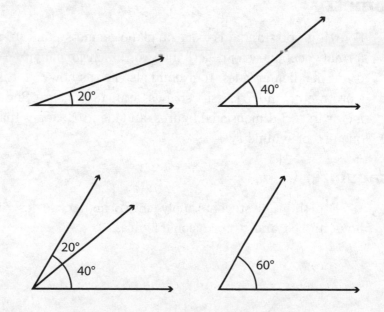

When a 20-degree angle is placed adjacent to a 40-degree angle, the measure of the outside angle is 60 degrees. Angle measures can also be decomposed, which is often part of a strategy for solving angle-measure problems.

EXAMPLE 26

Find the measure of angle *P*.

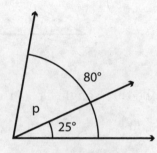

Discussion

The measure of the outer angle is 80 degrees. The sum of p degrees and 25 degrees is 80 degrees.

$$p + 25 = 80; p \text{ must be } 55$$

Geometry

Fourth-grade standards focus on plane geometry, and students begin to use formal vocabulary for describing two-dimensional figures as having parallel or perpendicular sides. Your child also learns how to measure angles using a protractor, and to classify shapes by angle measure. She will find line symmetry in two-dimensional figures, and identify shapes that have more than one line of symmetry.

Vocabulary

Fourth-grade students apply new vocabulary to describe and categorize lines, angles, and 2-dimensional figures.

LINES

- **Line:** A straight path connecting two points and extending beyond the points in both directions.
- **Line Segment:** A part of a line that has two endpoints.
- **Ray:** A half of a line that has one endpoint, and extends forever in the opposite direction.
- **Parallel Lines:** Lines in the same plane that never cross.
- **Intersecting Lines:** Lines in the same plane that cross at a point.
- **Perpendicular Lines:** Lines that form a right angle where the lines intersect.

ANGLES

- **Right Angle:** An angle whose measure is exactly 90 degrees.
- **Acute Angle:** An angle whose measure is less than 90 degrees.
- **Obtuse Angle:** An angle whose measure is more than 90 degrees and less than 180 degrees.

3-SIDED FIGURES

- **Right Triangle:** A triangle that contains a right angle.
- **Obtuse Triangle:** A triangle that contains an obtuse angle.
- **Acute Angle:** A triangle whose largest angle is less than 90 degrees.

4-SIDED FIGURES

- **Quadrilateral:** A closed two-dimensional figure with four sides that are line segments.
- **Parallelogram:** A quadrilateral where opposite sides are parallel and the same length.
- **Trapezoid:** A quadrilateral with exactly one pair of parallel sides.
- **Rectangle:** A quadrilateral with four right angles.
- **Rhombus:** A quadrilateral with four congruent sides.
- **Square:** A rhombus with four right angles.

EXAMPLE 27

Match each picture with a geometry term.

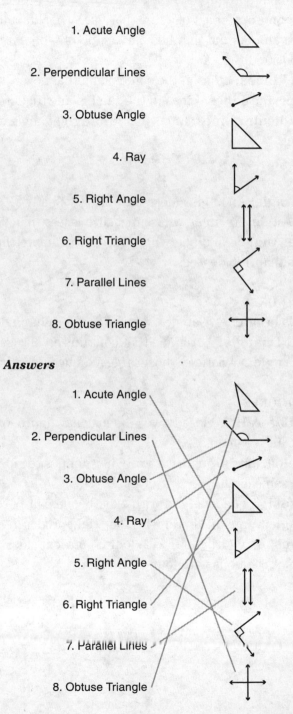

1. Acute Angle

2. Perpendicular Lines

3. Obtuse Angle

4. Ray

5. Right Angle

6. Right Triangle

7. Parallel Lines

8. Obtuse Triangle

Answers

1. Acute Angle

2. Perpendicular Lines

3. Obtuse Angle

4. Ray

5. Right Angle

6. Right Triangle

7. Parallel Lines

8. Obtuse Triangle

Symmetry

Fourth graders learn to understand symmetry in terms of *line symmetry*, where a two-dimensional figure can be folded along a straight line, which creates two halves that match exactly. They examine geometric figures for line symmetry as well as objects in the environment, letters of the alphabet, and designs of their own creation.

EXAMPLE 28

Identify which shapes have no lines of symmetry, one line of symmetry, or more than one line of symmetry.

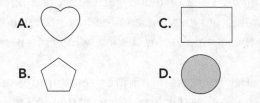

Discussion

The heart has one line of symmetry, a vertical line than runs down the middle. The pentagon (B) has five lines of symmetry: It can be folded along a line that begins at any corner and extends to the middle of the opposite side. Students often look at shapes and think of symmetry in terms of lines that go from the middle of one side to the middle of the opposite side, or from one corner to an opposite corner, and miss the lines of symmetry that run from sides to corners. The rectangle (C) has horizontal and vertical lines of symmetry. The question asks to identify the shapes as having zero, one, or more than one line of symmetry; it was not required to count the total number of lines of symmetry. The circle has infinite lines of symmetry.

Helping Your Child Succeed

Many topics in grade 4 will be new to your child, such as multiplying with fractions, performing calculations with larger numbers, using decimals to represent $\frac{1}{10}$ and $\frac{1}{100}$, measuring angles, and converting units of measure. You can assist your child by helping with homework and making connections between fractions and your daily routine.

Monitoring Your Child's Progress

Early in the year as you begin your journey through fourth grade, you will receive your child's assessment test results from the end of grade 3 (either PARCC or Smarter Balance, or your state's assessment). Look for areas of strength and weakness. The questions should be identified by the standard domains. Develop a game plan that could include your child's teacher to make sure deficits are addressed.

Around the House

Supplies for home use that will help your child succeed in fourth grade include a ruler, a meter stick or meter/yard stick combination, a protractor, plastic sheet protectors, dry erase markers, and a good supply of index cards. You can use these supplies for the following activities:

- The meter stick and protractor are great for measuring around the house, but the meter stick is particularly helpful for acting as a giant number line since it is marked off from 1 to 100. This can be helpful for modeling and solving homework problems with a number line to provide a quick visual model. Use the meter stick to help your child understand $\frac{1}{10}, \frac{2}{10} \ldots \frac{10}{10}, \frac{1}{100} \ldots \frac{100}{100}$ and compare 0.70 to 0.07.

- Help your child measure things that are longer than one meter or one yard to work on fractional lengths. Challenge your child to identify objects that are within a range of lengths or heights that you specify, such as between 48 and 65 inches.

- Estimate the angles at the top of a house. Find examples of right, acute, and obtuse angles. Which is most common? Are they standard?

- Have your child identify parallel and perpendicular lines around the house. Is there a reason why the lines are parallel or perpendicular? Have your child measure the angles of everyday objects. What is a good angle for a doorstop? What angle would not be good to use?

- Create a chart and collect 100 data points over the course of 2 weeks, for any data that is of interest to your child. Have her use the data to create a graph. What information can she extract from her graph?

- Use plastic sheet protectors and dry erase markers to create cool sets of reusable activities. Find templates of fraction circles and fraction blocks online and print them out. Putting the printouts into sheet protectors (or laminating them) makes them reusable and they can be marked up with dry erase markers. Templates such as this one can be good for helping with homework or modeling problems around the house.

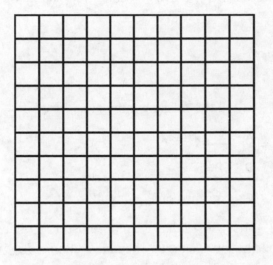

- Use index cards to create a set of cards containing representations of 1, $\frac{1}{2}$, $\frac{1}{3}$, $\frac{1}{4}$, and $\frac{1}{8}$. Make 8 copies of each one. Have your child practice adding and subtracting these fractions. Ask her to demonstrate $5 \times \frac{1}{8}$, and practice solving fraction addition problems such as $2\frac{1}{3} + 3\frac{2}{3}$ or $\frac{2}{3} - \frac{1}{3}$.

Around the World

Your computer can provide you access to valuable resources. There are math games, math references, and math practice problems available online, but there are also other great things around the world that are not necessarily math-centered, but can nurture your child's math awareness.

Think about what topics are of interest to your child, and what organizations would collect data that is related to it. For example, there are always monstrous engineering projects going on around the world. Where is the tallest building currently under construction? Will it be the tallest building in the world? If so, by how much; if not, by how much will it miss that designation? What airplane can carry the most number of people? How fast does a rocket have to go to get into orbit? What is the average population of the fifty states? Do you live in a state that has more, or less, than the average?

Exercises

1. Fill in the missing numbers:

   A. _____ $\times 8 = 56$

   B. $51 = 17 \times$ _____

   C. 2 groups of _____ are 62.

2. Miss Curtis has 12 more posters in her classroom than Miss Henderson. Together they have 28 posters. Write an equation or draw a diagram to model this problem, using p to represent the number of posters that Miss Curtis has.

3. Larry spent $55 for 5 baseball hats. How much did each hat cost?

4. Complete the patterns by filling in the missing numbers:

A.	4	11	____	____	284	851	
B.	3	8	18	38	____	158	318	____
C.	5	8	14	26	50	98	____	____

5. Write the number that is 10 times as great as the number given.

   A. 845

   B. 16,831

   C. 172

   D. 4,180

6. Round each number to the place value indicated.

   A. 14,831 round to nearest hundred

    B. 168 round to nearest ten

    C. 87 round to nearest thousand

    D. 145,876 round to nearest thousand

7. Solve

$$260 \div 9 =$$

8. Solve

$$42 \times 51 =$$

9. Tina read $\frac{1}{2}$ page of her book in the morning, and $4\frac{1}{2}$ pages at night. How many pages did she read in total that day?

10. Mrs. Kelly bought enough hamburger to make each of her 24 students a quarter-pound hamburger. How much total hamburger did she buy?

11. Represent $\frac{46}{100}$ using the visual fraction model.

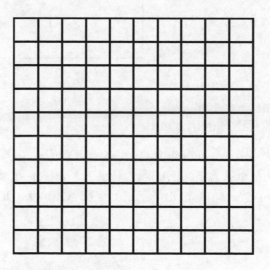

12. Compare the fractions using >, =, or <.

    A. $\frac{3}{4}$ ___ $\frac{3}{5}$ **B.** $\frac{5}{10}$ ___ $\frac{5}{100}$ **C.** 0.03 ___ 0.3

13. Peggy's bedroom has a perimeter of 108 feet. The length of the bed-room is 9 feet.

    A. What is the width of Peggy's bedroom?

    B. What is the area of Peggy's bedroom? Explain your answer.

14. Sally measured her dining room table to be $3\frac{1}{2}$ feet long. How many inches long is the table?

15. Anita's hair was $6\frac{1}{2}$ inches long in January and grew $3\frac{1}{2}$ inches by September. How long was Anita's hair in September?

16. The hands of a clock at 1:00 form a 30-degree angle. The angle increases by 30 degrees each hour. What is the angle of the hands of a clock at 3:00, 5:00, and 8:00?

17. Select all the words that describe this shape:

- ☐ Square
- ☐ Rhombus
- ☐ Quadrilateral
- ☐ Rectangle
- ☐ Trapezoid

18. Draw a rhombus that does not have any right angles.

19. Draw a quadrilateral that is both a rectangle and a rhombus.

20. Draw and count the lines of symmetry for this hexagon:

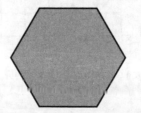

CHAPTER 9

Grade 5

Welcome to fifth grade! Grade 5 will extend the operations that students perform with fractions. While fourth graders can add and subtract fractions that have the same denominator and can find equivalent fractions, fifth-grade students add, subtract, and multiply fractions whose denominators are not necessarily the same. They will also use unit fractions while learning to divide fractions. They will represent fractions as decimals, and perform operations on numbers containing up to two decimal numbers. Students will also learn that volume Is another attribute of three-dimensional objects.

What Your Fifth Grader Is Expected to Learn

In this year, your child will expand on her vast knowledge of math concepts like place value and fractions, and delve into new concepts such as mixed numbers. In fifth grade your child will learn to:

- Extend her understanding of place value to include decimal numbers
- Multiply and divide fractions
- Apply her experience finding equivalent fractions from fourth grade to add and subtract fractions that don't necessarily have the same denominator
- Perform calculations with mixed numbers
- Use area models, arrays, number lines, drawings, and equations to explain division and mixed-operation problems
- Simplify expressions that use parentheses and other grouping symbols, and translate verbal models to numeric expressions and equations.
- Plot ordered pairs on a coordinate plane and solve problems using line plots
- Recognize volume and different ways to specify how much room something takes up

Common Core State Standards: Grade 5

There are five general categories of math skills your child will be learning in fifth grade and certain abilities that fall under each of those categories:

OPERATIONS AND ALGEBRAIC THINKING
- Write and interpret numerical expressions.
- Analyze patterns and relationships.

NUMBER AND OPERATIONS IN BASE TEN
- Understand the place value system.
- Perform operations with multi-digit whole numbers and with decimals to hundredths.

NUMBER AND OPERATIONS—FRACTIONS
- Use equivalent fractions as a strategy to add and subtract fractions.
- Apply and extend previous understandings of multiplication and division to multiply and divide fractions.

MEASUREMENT AND DATA
- Convert like measurement units within a given measurement system.
- Represent and interpret data.
- Geometric measurement: Understand concepts of volume and relate volume to multiplication and to addition.

GEOMETRY
- Graph points on the coordinate plane to solve real-world and mathematical problems.
- Classify two-dimensional figures into categories based on their properties.

For the full list of Common Core State Standards for Mathematics, see Appendix A.

What Your Child Should Know Before Fifth Grade

Students entering fifth grade should be able to count, add, and subtract using whole numbers up to 1 million; they should be able to multiply and divide two-digit numbers. If your child isn't fluent multiplying single-digit numbers, he will likely struggle in fifth grade. It's a good idea to take some time over the summer to work on becoming proficient with multiplication.

Your child should have a basic understanding of fractions. He should understand the numerator of a fraction as the number of parts of a whole, partitioned into the number specified by the denominator. He should be able to find equivalent fractions by multiplying the numerator and denominator by the same number. Students entering fifth grade should be able to add, subtract, and compare fractions when the denominators are the same. Your child should be able to multiply a fraction by a whole number, hopefully connecting that to adding the same fraction a specified number of times. For example:

$$3 \times \frac{1}{8} = \frac{1}{8} + \frac{1}{8} + \frac{1}{8} = \frac{3}{8}$$

In fifth grade, students continue to plot and compare numbers on a number line, and use the symbols <, =, and > to compare whole numbers, fractions, and decimals down to hundredths. Students should be able to convert fractions that have a denominator of 10 or 100 into decimals; they should see $\frac{3}{10} = 0.3$ and $\frac{26}{100} = 0.26$.

Operations and Algebraic Thinking

In fifth grade, your child writes and evaluates expressions containing multiple operations and will be able to write expressions from verbal models. For example, the perimeter of a rectangle can be found by finding the sum of twice the length and twice the width: $2 \times l + 2 \times w$, or by finding two times the sum of the length and the width: $2 \times (l + w)$. Expressions using grouping symbols, such as parentheses, are used to alter the normal order in which operations are evaluated. Students use their understanding of multi-operation expressions to explore the relationship between two rules, or expressions.

Using Grouping Symbols

When grouping symbols are used in an equation, such as parentheses, the expression inside the symbols must be performed first. This can help make expressions that have multiple operations easier to read. Grouping symbols can also alter the order in which operations would otherwise have been performed.

EXAMPLE 1

Add parentheses to the expression to show how it should be evaluated.

$2 + 3 \times 4 + 3$

Discussion

In the normal order of operations, multiplication and division operations are always done before addition and subtraction. Using parentheses helps to make this clearer, so the example becomes $2 + (3 \times 4) + 3$. After evaluating the expression inside the parentheses, this becomes $2 + 12 + 3$, which is 17.

Here the parentheses don't change the order in which the operations would have been performed; it just makes the expression more clear.

Parentheses can be used to group numbers and operations together to change the order in which operations are performed. When there are parentheses in an expression, first evaluate the expression inside the parentheses; then perform all of the multiplication and division operations, working from left to right; then perform all of the addition and subtraction operations, working from left to right.

EXAMPLE 2

Add parentheses to the expression $2+3\times4+3$ so it evaluates to 35.

Discussion

Multiplication would ordinarily be performed first, so try adding parentheses to force the addition operations to be performed first.

$$(2+3)\times4+3=5\times4+3=23$$

$$2+3\times(4+3)=2+3\times7=23$$

$$(2+3)\times(4+3)=5\times7=35$$

For more complex expressions using multiple sets of parentheses such as $(((2+3)\times4+5)\times6)$, evaluate the innermost set of parentheses first. To avoid using multiple sets of parentheses, other grouping symbols are used, in particular [ ] and { }. The previous expression could be written $\{[(2+3)\times4+5]\times6\}$.

QUESTION

Are there other grouping symbols?
The division bar is also a grouping symbol.

$$\frac{3\times2\times5-5}{2\times2+1}=\frac{30-5}{4+1}=\frac{25}{5}=5$$

In this example, the expression on the top should be evaluated; then the expression on the bottom should be evaluated; then the division operation should be performed. Parentheses are implied around the numerator and around the denominator.

EXAMPLE 3

Ricky and Jenny are working on the following homework problem: Evaluate the expression $14 - 3 \times 3 - 1$. Ricky says the answer is 22, and Jenny says the answer is 32. Who do you agree with? Explain your answer.

Discussion

The order of operations says to do all of the multiplication and division first, working from left to right. Multiply 3×3. The problem then becomes $14 - 9 - 1$. Now do all addition and subtraction operations, from left to right: $14 - 9$ is 5, and $5 - 1$ is 4. The answer is 4, and neither Ricky nor Jenny had the correct answer. A common mistake is to perform all operations from left to right. An incorrect answer would be found by starting with 14 and subtracting 3 to get 11, then multiplying 11 by 3 to get 33, then subtracting 1 to get 32, like Jenny's answer.

ALERT

When evaluating expressions that have multiplication or division mixed with addition or subtraction, add parentheses before you start. This will help to group the operations that should be performed first. For example. rewrite the expression $14 - 3 \times 3 - 1$ as $14 - (3 \times 3) - 1$. Remember to always simplify what is inside parentheses first.

EXAMPLE 4

Tommy is saving money from his part-time job so he can buy a new laptop that costs $450.00. He has already saved $100.00, and he just got a coupon in the mail worth $50.00 toward a laptop. He thought it would take 3 months for him to earn the rest of the money he needed, but his aunt said she would pay half of the money he still needed. Which of the following expressions represents the amount of money that Tommy still needs to save?

A. $450 - 100 - 50 \div 2$

C. $(450 - 100 - 50) \div 2$

B. $450 \quad (100 - 50) \div 2$

D. $450 \div 2 - 100 - 50$

Discussion

The cost of the laptop is $450. Tommy has already saved $100 and he has a coupon for $50, so the amount of money Tommy still needs is $(450 - 100 - 50)$. Since his aunt said she would pay half of the money that he still needs, she will pay half of $(450 - 100 - 50)$ and Tommy will need to save the other half.

EXAMPLE 5

You have homework problems to do that use large numbers. Your teacher says it's okay to use a calculator, but you forgot yours at school. You call someone with a calculator to get the answer to the expression $7 \times (14{,}981 + 876)$. What are the words that you use on the phone to make sure the person you are calling performs the calculation correctly?

Discussion

The parentheses in the problem direct you to perform the $14{,}981 + 876$ calculation first. The problem then becomes $7 \times (15{,}857)$.

Patterns and Relationships

When students perform the same calculations over and over that involve the same operations but different numbers, they may identify a pattern. They may see relationships between the values on the x-axis of a graph and the values on the y-axis, or between the x and y coordinates of values plotted on a graph. They may also see relationships between lists of numbers in a table. They may be able to define a rule that expresses the relationship between two columns in a table. Students learn to take values from a table and represent them in a graph, or take values from a graph and represent them in a table. Once students gain comfort in representing relationships in a variety of formats, they use those skills to analyze how two rules or graphs are similar or different.

EXAMPLE 6

Complete the following table, where the rule is $2 \times x + 3 = y$.

x	y
0	3
1	5
2	
4	11
	15
9	

Discussion

In class your child has probably encountered variations of "what's my rule?" games, where she sees a table of data and needs to figure out what the rule is. Here the rule is given in a way that the x and y values could be plotted on a graph, if desired. The first two lines are complete; the next line asks to find the output when the input is 2, which is found using $2 \times 2 + 3 = 7$. Now your child may see a pattern in the right column, 3, 5, 7; she may see that a 9 is missing but then the pattern picks up again with 11, or notice that in the left column the value for 3 is absent. She is likely to think about the patterns and complete the table with rows that are not being asked for. This will help her recognize the missing numbers in the left column. As a list of *ordered pairs* the values in the table are (0, 3), (1, 5), (2, 7), (4, 11) (6, 15), (9, 21).

EXAMPLE 7

Samantha and Kim experimented with helium balloons to see how fast the balloons rise. They wanted to know how high a balloon would rise after 1 second, 2 seconds, and up to 5 seconds after they let go of the balloon. They recorded the average times in the following table.

Time (seconds)	Height (feet)
1	10
2	20
3	30
4	40
5	50

Which of the following graphs best represents Samantha and Kim's data?

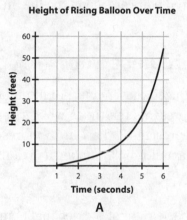

A

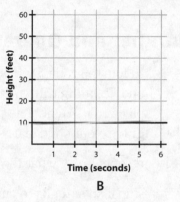

B

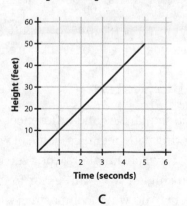

C

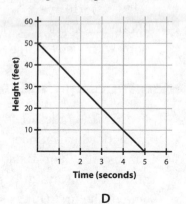

D

Discussion

The chart shows that after 1 second, the balloon is 10 feet off the ground. Graph A starts very slowly, and even after one second it is still on the ground; the height on graph A doesn't reach 10 feet until after 4 seconds, and then the graph increases rapidly. The table says the balloon should reach a height of 40 feet after 4 seconds, but graph A shows a height of only 10 feet. Graph B shows a constant height at 10 feet, so this does not represent an object whose height is increasing. Graph D shows an object that begins at a height of 50 feet, and *decreases* ten feet every second, so this is a graph of something that is falling. Graph C represents the data. If you take the data in the table and create ordered pairs from it (1, 10), (2, 20), (3, 30), (4, 40), and (5, 50) you will see that each of these ordered pairs are along the line in graph C.

Number and Operations in Base Ten

In fourth grade, students recognized that in multi-digit numbers such as 17 and 75, a digit in one place represents ten times what it represents in the place to its right. The 7 in 17 represents 7, the 7 in 75 represents 70. In fifth grade, your child learns that in a multi-digit number, a digit in one place represents $\frac{1}{10}$ of what the number represents to its left—and that this relationship also extends to numbers to the right of the decimal point. The 2 in 0.25 represents $\frac{2}{10}$, while the 2 in 0.62 represents $\frac{2}{100}$. Students in fifth grade use exponents applied to a base of 10 for the first time. Students perform operations with whole numbers and decimals down to the hundreds place.

Understanding the Place Value System

In fifth grade, students notice similarities and differences between working with whole numbers and working with decimals, some of which will cause confusion. When students compare whole numbers, they see that a five-digit number is always greater than a single-, two-, three-, or four-digit number. One of the first things students may do when comparing whole numbers is to compare the length of the string of numbers. This method will not always work when comparing decimal numbers. For example, 0.5 is greater than

0.15, and 0.15 is greater than 0.149. One method for comparing decimals is to adjust all numbers to have the same number of digits to the right of the decimal point by adding zeros to the end of the number, such as 0.50 and 0.150.

Base Ten Exponents

Fifth-grade students use exponents for the first time, using 10 as the base. They represent 10×10 as 10^2 and $10 \times 10 \times 10$ as 10^3. They relate the number of zeros to the number of the exponent on the 10. In later grades, this will help them with scientific notation and using powers with bases other than 10.

EXAMPLE 8

Which number is equal to 10^5?

A. 100   **C.** 100,000

B. 10,000   **D.** 1,000,000

Discussion

Students approach this problem from many directions. One student may write 10^5 as a 1 with five zeros (100,000). Others may convert each of the four choices to 10^2, 10^4, 10^5, and 10^6, then choose the answer that matches the question. Yet another student may start with a power of 10 that he is most familiar with, for example $10^3 = 1000$, and work up to 10^5.

EXAMPLE 9

Use mental math to multiply or divide to evaluate the following expressions.

A. The number of eggs in ten dozen: 12×10

B. The number of $10 bills needed to have $1,500: $1500 \div 10$

C. The amount of money raised, if 1,200 tickets were sold for a school play at $10 each: $1{,}250 \times 10$

D. The number of penny rolls needed to wrap 10,000 pennies in rolls of 50: $10{,}000 \div 50$

Discussion

The answers for A though C can be found by adding or removing one zero to the end of the number. D is slightly different, and students can view it in different ways. One student may calculate that there are 100 pennies per roll, divide 10,000 by 100, and then multiply that by 2 because there are only 50 pennies per roll, not 100. Another student may divide 10,000 by 10 to get 1,000, and then divide by 5 to get 200.

In the *base ten* number system, each digit has ten times the value of the same digit one place to the right: $444 = 400 + 40 + 4$. This is true for all digits of a decimal number, too: $555.55 = 500 + 50 + 5 + 0.5 + 0.05$. Conversely, each digit has $\frac{1}{10}$ the value of the same digit one place to the left.

Place Value with Decimals

In fifth grade, students begin to represent fractions with decimal numbers. Many states have incorporated money standards into their adaptation of the Common Core Standards. Using money problems provides students with real-world fraction and decimal applications. Students can relate $\frac{1}{4}$ of a dollar as twenty-five cents. They know it is written to the right of a decimal point because it is worth less than $1. Students would write a quarter as .25. Using a leading zero with decimal numbers, such as 0.25, emphasizes the decimal, helps students see that the number is less than one, and makes it easier to read.

The number representing $1,325.49 is added to the following place value chart.

Ten Thousands	Thousands	Hundreds	Tens	Ones	[.]	Tenths	Hundredths
10,000	1,000	100	10	1		$\frac{1}{10}$	$\frac{1}{100}$
	1	3	2	5	.	.4	9

The number 1 represents 1,000, the number 3 represents 300, the number 2 represents 20, the number 5 represents 5 (or 5 ones), the number 4 represents the number of tenths (or 0.40), and the number 9 represents the number of one-hundredths or 0.09. Fifth graders learn to use decimal numbers into the $\frac{1}{1000}$ position. They will see patterns, such as multiplying by 10 adds a zero to the right of a whole number, or moving the decimal point one place to the right; they will see that dividing by 10 is the same as moving the decimal point one position to the left.

Here are some examples of place value questions that your fifth grader may be expected to answer:

What is the value of the 3 in the number 4.231?

Fifth-grade students learning place value with decimal digits may choose to model the problem using a place value table. For example, here is a model of the number 4.231:

Tens	Ones	[.]	Tenths	Hundredths	Thousandths
10	1		$\frac{1}{10}$	$\frac{1}{100}$	$\frac{1}{1000}$
	4	.	2	3	1

The digit 3 is in the $\frac{1}{100}$ place. The answer is that the value of 3 in the number 4.231 represents 3 one-hundredths, or $\frac{3}{100}$.

Your child might choose to model the problem using a place value chart and write the number in expanded form:

$$(4 \times 1) + (2 \times \frac{1}{10}) + (3 \times \frac{1}{100}) + (1 \times \frac{1}{1000})$$

First, students write the number representing the part to the left of the decimal point, then they use the word *and* to indicate that there is a fractional part of the number. They then find the smallest unit that has a value—here there is a 1 in the thousandths place, so the decimal part of the number will be expressed in thousandths. 4.231 is written as "four and two hundred thirty one thousandths."

Note that 0.2 would be written as two tenths, but 0.200 would be written as two hundred thousandths. The numbers represent the same value, but if the extra zeros are added for precision then the verbal model should be just as precise.

EXAMPLE 10

Which point on the number line best represents the location of 3.6?

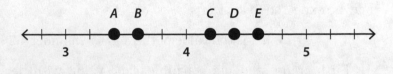

Discussion

Students first narrow down the choices to A or B because they will recognize that 3.6 is between 3 and 4. Next they have to see what each of the lines between the 3 and the 4 represent (these lines are called *pips*). They will see that each pip represents 0.2 because each distance between the whole numbers is partitioned into 5 units, so each unit must represent two-tenths. A represents 3.4, and B represents 3.6.

EXAMPLE 11

In April 2014, two runners at the Drake Relays (Brianna Rollins and Kristi Castlin) ran the 100-meter hurdles in times of 12.576 and 12.571 seconds. What was the winning time?

Discussion

Order, sort, and compare the numbers down to the one-thousandth place. Strategies can include using a number line, or a place value chart. Another is to read the numbers from left to right until the digits no longer match. Reading from left to right, you will see that the numbers left of the decimal point are the same (12). Continue to read from left to right, checking the numbers to the right of the decimal point. Both times have a 5 in the tenths digit, so the digits in the hundredths place are then compared. They are both 7. Then the digits in the thousandths place are compared—one is

6 and the other is 1, so they are different. Now that the difference has been identified, remember that you are looking for a winner of a race, so you are looking for the *lowest* time. The winner of the race had the time of 12.571 seconds.

EXAMPLE 12

Sort the following batting averages of the New York Yankees into two boxes, one with the averages less than .270 and one with averages at least .270. Then in each box, order the batting averages from lowest to highest.

.271, .258, .255, .285, .235, .217, .233, .237, .219, .254

Discussion

For the first pass through the data, compare each average to .270 to see which box they belong in. Averages that are less than .270 include .258, .255, .235, .217, .233, .237, .219, .254. Averages at least .270 include .271, .285. Next, put the numbers from each box in order from lowest to highest.

Averages below .270	Averages at least .270
.217	.271
.219	.285
.233	
.235	
.237	
.254	
.255	
.258	

EXAMPLE 13

Jake bought 3 bananas at $0.38 per pound. They weighed a total of 2.25 pounds. The total cost should have been $0.855, but the store needs to round to the nearest cent. How much did Jake pay for bananas?

Discussion

In this problem the student is asked to round to the nearest cent, which is the $\frac{1}{100}$ place. To round to the nearest $\frac{1}{100}$ place, look at the digit to its right—in the $\frac{1}{1000}$ place. Because the digit in the $\frac{1}{1000}$ place is 5 or higher, the store will round up from $0.855 up to $0.86.

Operations with Whole Numbers and Decimals

A big concern for parents about the Common Core Standards is that their children won't learn the standard algorithms they learned in school. The standards guide students through illustrating and explaining calculations using equations, tape diagrams, rectangular arrays, and area models, so that students do develop a deep understanding of the standard algorithms. In fifth grade, students become fluent with multiplying whole numbers using the standard algorithm. They use strategies based on place value and the properties of operations to learn about decimal numbers, and continue developing division skills.

Multi-Digit Multiplication

Deep understanding for the standard multiplication algorithm is built using place value, having students proficient with multiplying two single-digit numbers, and multiplying numbers by powers of 10. Students who understand the area model of multiplication will see the same calculations performed using the standard algorithm. For example, to calculate the number of inches in one mile, multiply the number of feet in one mile (5,280) by the number of inches in one foot (12).

$$5,280 \times 12$$

$$(5,000 + 200 + 80) \times (10 + 2)$$

	5000	200	80
10	50,000	2,000	800
2	10,000	400	160

After performing the multiplication calculation for each area, students find the sum of all six products or can add the values of each row $52,800 + 10,560$ or add the values of the columns $60,000 + 2,400 + 960$ to find there are 63,360 inches in a mile. Here is how the area model connects to the standard algorithm:

Area Model

$$5,280$$
$$\times 12$$

0	$2 \times 0 = 0$
160	$2 \times 80 = 160$
400	$2 \times 200 = 400$
10,000	$2 \times 5,000 = 10,000$
0	$10 \times 0 = 0$
800	$10 \times 80 = 800$
2,000	$10 \times 200 = 2,000$
50,000	$10 \times 5,000 = 50,000$
63,360	sum of all products

Standard Algorithm

$$5,280$$
$$\times 12$$

10,560	$2 \times 5,280$
52,800	$10 \times 5,280$
63,360	sum of both

Multi-Digit Division

Fifth-grade students become fluent multiplying numbers by 10, 100, and 1,000. Students can often divide products of 10, 100, and 1,000 by 2 using mental division, and they can multiply most two-digit numbers by 2 easily. They apply these skills to divide four-digit numbers by two-digit numbers.

EXAMPLE 14

Sara is treasurer of the Adventure Club at school. She found out that it will cost $9,180 to hire two buses for sixty students to take a five-day trip to visit national parks in Utah and Arizona. How much money will each student have to pay?

Discussion

Here is one way to arrive at a solution to the problem. Sara will divide 9,180 by 60. So first Sara makes a chart using multiples of 10:

If each student paid $1 it raises $60 total

For each $10 all students pay, it raises $600 total

For each $100 all students pay, it raises $6,000 total

From these calculations she added these lines to her table:

For each $2 all students pay, it raises $120.

For each $5 all students pay, it raises $300 total

For each $50 all students pays it raised $3,000 total

Now Sara thinks she is ready to divide 9,180 by 60.

If everyone pays $100, that raises 6,000, and leaves $9,180 - 6,000 = $3,180$

If everyone pays $50 more, that raises 3,000, and leaves $3,180 - 3,000 = 180

If everyone pays $2 more, that raises 120, and leaves $180 - 120 = 60

If everyone pays $1 more, that raises 60, and leaves $60 - 60 = 0$

So everyone has to pay $(100 + 50 + 2 + 1) = 153

This method of division represents repeated subtraction. The $9,180 must be distributed among 60 students. Every $1 distributed to each student reduces the amount left to distribute by $60. Rather than reducing the amount $60 at a time, larger multiples of 60 are used. Your child is learning to build charts like this with numbers she can calculate easily. She can use powers of 10, double numbers to get the multiple of 2, and get the multiples of 5 and 50 by dividing the products of 100 and 10 in half.

Modeling Decimal Division

When students understand division problems as repeated subtraction problems, they are able to extend their understanding to include decimal numbers. The following example walks through each step of a division problem that uses place value and properties of operations.

EXAMPLE 15

You and five friends are at a fair. You have $4.80, and would like to divide your money evenly so that everyone may buy a snack. Including yourself, how much money should each person get?

Discussion

The problem to be solved is $4.80 \div 6 = ?$

Write the division problem.	$6)\overline{4.80}$
Is there enough to give everyone 1.0? No, you would need to have 6.0. Is there enough to give everyone 0.5? Yes, you would need to have 3.0. Keep track that you gave everyone 0.50.	$6)\overline{4.80}$ $-\,3.00$ 1.80
You spent 3.00 and have 1.80 left. If you give everyone 0.10, it costs 0.60. If you give everyone 0.30 it costs 1.80, which is what you have left Keep track that you gave everyone 0.30.	$6)\overline{4.80}$ $-\,3.00$ 1.80 -1.80 0
Add up what you gave everyone.	$0.50 + 0.30 = 0.80$

Suppose that the money to be distributed was 480 pennies. The time-consuming method for sharing the money would be to pass it out one penny at a time. By applying place value, multiples of 10 and multiples of 6 (because there are 6 people to distribute money to), more efficient methods of distributing the money can be found. By understanding how to model division problems in this way, students have a deeper understanding for using standard algorithms.

Number and Operations—Fractions

In grade 4, students found equivalent fractions; in grade 5, students use equivalent fractions when they add, subtract, and maybe divide fractions with unlike denominators. They will be able to identify equivalent fractions and explain why two fractions are equal.

In grade 4, students understood larger fractions such as $\frac{3}{5}$ to be a combination of smaller fractions: $\frac{1}{5} + \frac{1}{5} + \frac{1}{5}$. In grade 5, students see this as $3 \times \frac{1}{5}$. They are able to multiply whole numbers and fractions, multiply fractions together, and multiply fractions and mixed numbers. In grade 5, students model dividing whole numbers by fractions, and fractions by whole numbers.

Later, in grade 6, students will use these skills to understand ratios and proportions, unit rates, and divide fractions with unlike denominators and mixed fractions.

Using Equivalent Fractions to Add and Subtract Fractions

When students add $\frac{1}{2} + \frac{1}{2}$ they see that two halves make one whole; they would not add $\frac{1}{2} + \frac{1}{2}$ and get $\frac{1}{4}$, because they see that when two halves of something are added they will not get a smaller piece. But when students add fractions that either have different numerators or different denominators, then all sorts of creative ways to get an incorrect answer are demonstrated. Numerically, students will get correct answers when they convert fractions to have a *common denominator*. The focus for adding and subtracting (and sometimes comparing) fractions should be on using common denominators; the Common Core Standard de-emphasizes finding the lowest (or least) common denominator. Students can always find a common denominator by finding the product of two denominators.

EXAMPLE 16

Bella built a house of cards $3\frac{3}{4}$ feet tall. Shelly's house of cards is $\frac{3}{8}$ feet taller. How tall was the house of cards Shelly built?

Discussion

Shelly's house is $\frac{3}{8}$ feet taller than $3\frac{3}{4}$ so the numbers should be added together.

One option is to covert $3\frac{3}{4}$ to eights, and then add 3 more eighths. The answer can also be viewed as adding $3+\frac{3}{4}+\frac{3}{8}$; the $\frac{3}{4}$ can be converted to $\frac{6}{8}$, so the answer can be found by adding $3+\frac{6}{8}+\frac{3}{8}=3\frac{9}{8}$. $\frac{9}{8}$ can be written as $1\frac{1}{8}$.

The problem can be rewritten $3\frac{9}{0}=3+1+\frac{1}{8}=4\frac{1}{8}$.

Connecting Adding Fractions to Multiplication

The following figure represents a model of $\frac{2}{3}$. It shows a whole divided into three pieces, with two of them shaded.

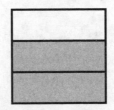

A model of $4 \times \frac{2}{3}$ shows the model of $\frac{2}{3}$ repeated 4 times.

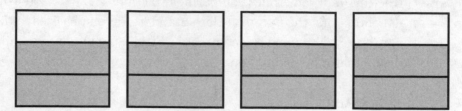

Each shaded region represents $\frac{2}{3}$. Adding up all of the shaded regions of the model, $4 \times \frac{2}{3}$ produces eight thirds, $\frac{8}{3}$. When any whole number and fraction are multiplied, the product will be the whole number times the numerator of the fraction, with the denominator of the fraction becoming the denominator in the product.

$$4 \times \frac{2}{3} = \frac{(4 \times 2)}{3} = \frac{8}{3}$$

This also models how multiplication is a form of addition, where $\frac{2}{3}$ is added 4 *times*:

$$\frac{2}{3} + \frac{2}{3} + \frac{2}{3} + \frac{2}{3} = \frac{8}{3}$$

ESSENTIAL

Students often find it easier to write whole numbers in fraction form, and then "multiply straight across," meaning multiply the numerators, and multiply the denominators.

$$4 \times \frac{2}{3} = \frac{4}{1} \times \frac{2}{3} = \frac{8}{3}$$

Mixed Numbers and Improper Fractions

When 4 is multiplied by $\frac{2}{3}$ the product is $\frac{8}{3}$. A model of $\frac{8}{3}$ is shown as follows; each shaded region represents $\frac{1}{3}$ of a whole. The total shaded region represents $2\frac{2}{3}$.

An *improper fraction* is one where the numerator is greater than (or equal to) the denominator, and the fraction represents a number has at least one whole. A *mixed number* has a whole number and a fraction.

A fraction is in simplest form when the numerator and denominator do not have any factors in common greater than 1. For example, $\frac{4}{6}$ is not in simplest form because 2 is a common factor to both 4 and 6. To simplify $\frac{4}{6}$, divide both the numerator and denominator by 2.

$$\frac{4}{6} = \frac{(4 \div 2)}{(6 \div 2)} = \frac{2}{3}$$

Dividing with Fractions

Most adults were taught as kids that when dividing by a fraction, you "flip" the divisor and multiply. Not much discussion was provided; it just works, and it works whether the dividend is a whole number or a fraction. For example:

$$4 \div \frac{2}{5} = 4 \times \frac{5}{2} = \frac{20}{2} = 10$$

Now, suppose that we don't flip and multiply. What if a common denominator was found, just as we do for addition and subtraction, and what if we divided straight across?

EXAMPLE 17

Solve: $4 \sqrt{\dfrac{2}{5}}$

Discussion

The common denominator for $\dfrac{4}{1}$ and $\dfrac{2}{5}$ is 5, so rewrite expressions with the common denominator :

First, rewrite 4 with a denominator of 5: $\dfrac{4}{1} \times \dfrac{5}{5} = \dfrac{20}{5}$

Next, rewrite the expression to: $\dfrac{20}{5} \sqrt{\dfrac{2}{5}}$

then divide straight across: $\dfrac{20}{5} \div \dfrac{2}{5} = \dfrac{20/2}{5/5} = \dfrac{10}{1} = 10$

Notice that the denominator $\dfrac{5}{5} = 1$, so the fraction simplifies to $\dfrac{10}{1}$. *Using the common denominator and dividing will always produce 1 in the denominator.* The net effect is the same as "flipping" the divisor and multiplying.

The Common Core approach encourages students to model division, even with fractions. Keep in mind that when any number (the dividend) is divided by any divisor, the question is asking, "How many times can the divisor be subtracted from the dividend?" Division problems with fractions can be modeled both by number line models and circle models.

EXAMPLE 18

Ashley was given the following problem:

Lucy has $\dfrac{1}{2}$ gallon of milk that she will pour into 4 same-sized containers.

How much milk will she have in each container? Explain your answer.

Discussion

Here is Ashley's solution:

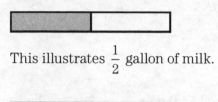

This illustrates $\frac{1}{2}$ gallon of milk.

The $\frac{1}{2}$ gallon of milk is divided into 4 equal amounts.

That means the $\frac{1}{2}$ gallon that is gone could have been divided into 4 equal amounts.

This is an example of a tape diagram showing that if $\frac{1}{2}$ of the gallon represents 4 equal amounts, then the whole gallon can be represented as 8 equal amounts. When Lucy divides the $\frac{1}{2}$ gallon into 4 same-sized containers, each container will have $\frac{1}{8}$ of a gallon of milk.

This models $\frac{1}{2} \div 4 = \frac{1}{8}$.

EXAMPLE 19

Marvin was given the following math problem:

Joey bought 4 small pizzas that were divided into 3 slices. The slices were small, so he wanted to serve each of his friends 2 slices. Including himself, how many people can Joey serve? Use the circle diagram to model $4\sqrt{\frac{2}{3}}$ and explain your answer.

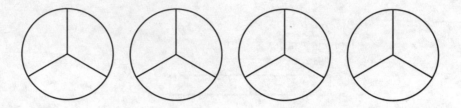

Discussion

Here is how Martin solved the problem: Each of Joey and his friends get $\frac{2}{3}$ of a pizza, so he labeled Joey's two pieces with the letter J, and counted $\frac{2}{3}$s for each friend. Friend #1 gets the pieces labeled F1, and the pieces are similarly labeled for friends 2, 3, 4, and 5.

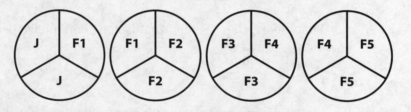

There are enough slices for Joey and five friends, so a total of 6 people can be served. This is a model of $4 \div \frac{2}{3} = 6$. Martin thinks his answer is reasonable because if everyone got a whole pizza there would be enough for only four people, but because everyone gets less than a full pizza there is pizza for more than 4 people. If everyone got $\frac{1}{2}$ of a pizza, there would be enough for eight people. Because everyone gets more than $\frac{1}{2}$, there isn't enough for 8 people.

One of the strategies Marvin used to reason that his answer was good is comparing fractions with $\frac{1}{2}$ and with one whole. Students will find this strategy helpful for addition, subtraction, and multiplication problems, as well as sorting fractions and plotting them on a number line. When students are asked to compare two fractions, it is often easier to compare each one with $\frac{1}{2}$

rather than each other. For example, when comparing $\frac{3}{8}$ to $\frac{9}{14}$, students may see $\frac{3}{8}$ is less than $\frac{1}{2}$ (or $\frac{4}{8}$) and $\frac{9}{14}$ is greater than $\frac{1}{2}$ (or $\frac{7}{14}$), so $\frac{3}{8} < \frac{9}{14}$.

Measurement and Data

Fifth-grade students convert units of measurement, such as feet to inches, to solve real-world problems. When measurements such as length are provided in feet and measurements for width are provided in inches, one (or both) of them need to be converted in order for the area to be found, or for the product of the two measures to have meaning. Measurement and Data standards also include representing and interpreting data from line plots, and geometric measurement standards to help students understand volume.

Grade 5 students encounter measures from two systems, the *metric system* and the system of *United States customary units* (or U.S. standard units), which has its roots in the English system of weights and measures in use at the time of American independence. (The English system was overhauled not long after to what is now called imperial units, adjusting measurements but keeping the same unit names; this can sometimes lead to confusion between the U.S. customary and imperial units.) The fifth grade introduces applying the units of measure to provide *scaling*—for example, moving from inches to feet, or feet to yards.

Converting units in grade 5 focuses on converting units of measure *within* the same system of measures: metric to metric conversion, and U.S. units to U.S. units. Later grades will convert metric to U.S. units, and U.S. units to metric.

The U.S. Customary Units of Measure

The units of length that fifth graders encounter are *inches, feet, yards,* and *miles*. When these length measurements are used to describe the perimeter of an area, the area is in units of *square inches, square feet, square yards,* or *square miles*. When these units of length are used to describe the dimensions of a three-dimensional container, the units of volume are *cubic inches*, *cubic feet*, or *cubic yards*, and (less frequently) *cubic miles*.

Liquid measures that fifth graders encounter include *ounces*, *pints*, *quarts*, and *gallons*. In applications such as baking, students may encounter *teaspoons*, *tablespoons*, and *cups*. Dry measures that fifth graders may encounter are *pints*, *quarts*, and occasionally *pecks* or *bushels*. During a trip to a farmers' market (or farm), students could encounter pints of blueberries, or bushels of apples. Common U.S. units of weight encountered by fifth graders include *ounces*, *pounds*, and *tons*.

▼ **U.S. UNITS OF MEASURE: LENGTH**

Unit	Units (plural)	Abbreviation	Common Conversions
inch	inches	in.	There are 12 inches in 1 foot; 36 inches in 1 yard.
foot	feet	ft.	There are 3 feet in 1 yard; 5,280 feet in 1 mile.
yard	yards	yd.	
mile	miles	mi.	

EXAMPLE 20

Ben is 5 feet 3 inches tall. Represent this height in inches.

Discussion

There are 12 inches in one foot. To convert 5 feet into inches, multiply 5 by 12. $5 \times 12 = 60$. Ben is 5 feet 3 inches tall, so he is $60 + 3$ inches tall, or 63 inches.

EXAMPLE 21

On Field Day during the long jump competition, Ben's jump measured 190 inches. Which of the following is another way to represent 190 inches?

A. 1 foot 9 inches   **C.** 15 feet 10 inches

B. 15 feet 8 inches   **D.** 19 feet

Discussion

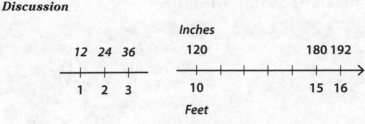

See how many whole units of feet Ben jumped. 190 is between 180 and 192, so there are 15 whole feet, which is 180 inches. Ben jumped 190 inches, which can be represented by 180 + 10 inches, and since 180 inches is 15 feet, the distance that Ben jumped can be represented by 15 feet 10 inches.

EXAMPLE 22

Mr. Duke put a rope around his garden that was 120 feet long. What is the length, in yards, of Mr. Duke's rope?

Discussion

Three feet = 1 yard. To convert from feet to yards, divide by 3, as there are 3 feet in each yard. Divide 120 yards by 3 to find the length of the rope in yards. $120 \div 3 = 40$. Mr. Duke's rope is 40 yards long. To check the answer, convert from yards back to feet. Multiply the number of yards by 3: 40 yards × 3 feet per yard = 120 feet.

The Metric System

The basic unit of mass in the metric system is the *kilogram*. Fifth-grade students are likely to encounter *grams* and *milligrams*. The metric unit of volume that fifth-grade students are most likely to encounter are *liters* and *milliliters*. A cubic centimeter has a volume equal to one milliliter, so a cube that is 10 centimeters long, 10 centimeters wide, and 10 centimeters high can hold one liter. Students who measure area using the metric system will use *square centimeters*, *square meters*, and occasionally *square kilometers*. Students measuring volume will use *cubic centimeters*, *cubic meters*, and occasionally *liters*.

▼ THE METRIC SYSTEM: LENGTH

Unit	Abbreviation	Common Conversions
Millimeter	mm	There are 10 millimeters in 1 centimeter, and 1,000 millimeters in one meter.
Centimeter	cm	There are 100 centimeters in 1 meter.
Meter	m	There are 1,000 meters in one kilometer.
Kilometer	km	

EXAMPLE 23

Mrs. Brooks gave each group of students 2 meters of duct tape to make a square on the floor of the classroom. What is the length of each side of the square, measured in centimeters?

Discussion

The units supplied by the question is *meters* but the units for the answer should be in *centimeters*. One approach would be to convert the units given (meters) to the units of the answer (centimeters). There are 100 centimeters in 1 meter, so multiply 100×2 to find the number of centimeters in 2 meters. $100 \times 2 = 200$, so Mrs. Brooks gave each group of students 200 cm of duct tape. There are four equal-side lengths of a square, so each side of the square is $200 \div 4 = 50$ cm.

ESSENTIAL

A meter stick is a nice item to have at home for a student. A meter stick may have metric units on one side, and inches and feet on the other.

Fifth-grade students may have the opportunity to find the mass of objects using a type of scale known as a triple beam balance. They would be able to measure using grams and milligrams.

▼ THE METRIC SYSTEM: MASS

Unit	Abbreviation	Common Conversions
milligram	mg	There are 1,000 milligrams in 1 gram.
gram	g or gr	There are 1000 grams in 1 kilogram.
kilogram	kg	

Fifth-grade students may have the opportunity to find the volume of liquids using beakers or graduated cylinders. They may calculate the volume of solid objects from the width, length, and height.

▼ **THE METRIC SYSTEM: VOLUME**

Unit	Abbreviation	Common Conversions
milliliter	ml	There are 1,000 milliliters in 1 liter.
liter	l	

A gram is the mass of 1 cubic centimeter of water. 1 cubic centimeter is also 1 milliliter. The volume of a 10 cm cube is $10 \times 10 \times 10 = 1,000$ cubic centimeters, which is 1 liter. 1 liter of water has a mass of 1 kg.

Representing and Interpreting Data

Students have represented and interpreted data in several formats prior to fifth grade, and will extend their use of familiar line graphs to represent fractional units of measure, particularly unit fractions.

EXAMPLE 24

Students from an elementary school in Vermont measured the snow that fell in their backyards during a snowstorm. This list shows the amount of snow measured by 10 of the students.

Jake	8.5	Ella	8.5
Aaron	10	Jackie	9.5
Marvin	10.5	Kelly	9
Sammy	10.5	Marie	10.5
Steve	8	Bella	9

Which of the following line plots correctly represents the snowfall measured by the students?

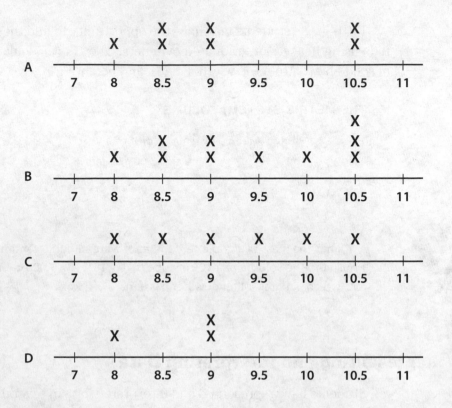

Discussion

One X represents one data point. Each of the 10 students contributes one X to the line graph, so there should be 10 Xs. Choice B correctly represents 10 of the data elements from the table, and it is the only line plot with 10 data points. Choice C represents each value of snowfall that occurred at least once, but does not represent measurements that were common to more than one student. Choices A and D having missing data. Can you identify the missing snow measures in those two line plots?

Geometric Measurement: Volume

Manipulatives, such as centimeter cubes, are a great way for students to understand volume and be able to relate volume to addition and multiplication. For example, if centimeter cubes are placed along a 10 cm length of an area, along a 5 cm width of the area, and then filled with $10 \times 5 = 50$ cubes, the volume of the area is 50 cubic centimeters. Adding another layer of cubes creates a volume of 100 cubic centimeters. Students should experience building shapes using unit cubes before using the formulas with volume.

EXAMPLE 25

In the metric system, one milliliter has the volume of one cubic centimeter, one liter is equivalent to 1,000 cubic centimeters. Noah wanted to build a model to represent $\frac{1}{4}$ liter, using centimeter cubes. Which of these models could Noah build to represent $\frac{1}{4}$ liter?

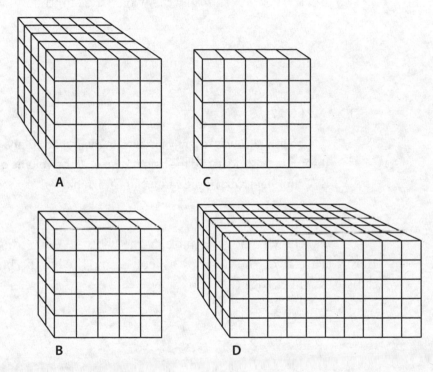

A   C

B   D

Discussion

One liter is equivalent to 1,000 cubic centimeters. Noah wants to build a model of $\frac{1}{4}$ liter, so the volume of Noah's model $= \frac{1}{4} \times 1000 = 250$ cubic centimeters. To calculate the volume, multiply the length $\times$ width $\times$ height.

The volume of figure A is $5 \times 5 \times 5 = 125$ cm^3.

The volume of figure B is $5 \times 2 \times 5 = 50$ cm^3.

The volume of figure C is $5 \times 5 \times 1 = 25$ cm^3.

The volume of figure D is $10 \times 5 \times 5 = 250$ cm^3.

EXAMPLE 26

A large fish tank was recently installed in the school cafeteria. The width of the tank is 2 feet, the length is 12 feet, and the height is 4 feet. Find the volume of the fish tank.

Discussion

To find the volume, multiply the width, the length, and the height.

$V = L \times W \times H.$

Volume $= 12 \times 2 \times 4 = 96$ cubic feet

EXAMPLE 27

A large fish tank was recently installed in the public library. The length of the fish tank is 8 feet, and the height is 3 feet. If the volume of the fish tank is 72 cubic feet, what is the width of the fish tank?

Discussion

To find the volume, you multiply the width, the length, and the height. Here you are given the volume, the length, and the height, but not the width. Substitute the values that you know.

$V = L \times W \times H$

$72 = 8 \times W \times 3$

This can be simplified by multiplying 8 and 3.

$72 = 24 \times W$

The value that makes this true is $W = 3$. The width of the fish tank is 3 feet.

Geometry

Reading and graphing points on a coordinate plane and solving and interpreting graphs of real-world math problems is new to fifth-grade geometry students. Students will continue developing their understanding of two-dimensional figures by categorizing shapes based on attributes such as the number of parallel lines, the number of right angles, and ways to distinguish shapes within broad categories such as quadrilaterals.

The Coordinate Plane

Grade 5 students are introduced to the *coordinate plane*, where the two axes represent a horizontal and vertical number line. They learn to plot points on the coordinate plane, and learn to identify the coordinates of a plotted point. The horizontal axis is called the *x-axis*, and the vertical axis is called the *y-axis*. They use a coordinate plane, changing the scale of the axis as appropriate, to represent the real world. *Coordinates* refer to the ordered pair of numbers (x, y) that represent the location of a point. The point $(0, 0)$ is where the two axes meet.

In the fifth grade, your child learns to create a graph and label the x-axis, the y-axis, and the *origin* (where the two axes meet or cross), then plot a point in the coordinate plane, such as $(4, 3)$. She can identify the coordinates (x, y) of a plotted point. She can create a graph on a coordinate plane to represent a real-world problem, using an appropriate scale on the axes and labeling them appropriately, then give the graph a title that reflects the data represented in it. She can also extract an ordered pair for a data point that is plotted in a coordinate plane, correctly interpreting the value represented by the x coordinate and the y coordinate.

EXAMPLE 28

Plot the point (4, 3) in a coordinate plane.

Discussion
Draw and label the x- and y-axes.

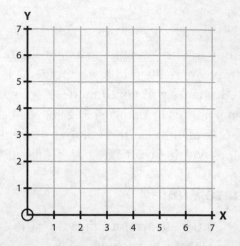

The first number (4) in the ordered pair (4, 3) represents the distance from the origin along the *x-axis*. Count four units from left to right along the bottom of the plane.

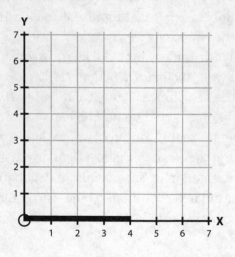

The second number (3) in the ordered pair (4, 3) represents how far up from the x-axis the point should be plotted. Use the scale on the *y-axis* to help count up 3 units.

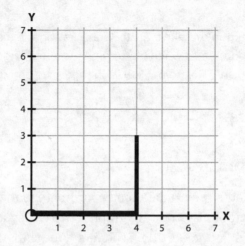

Draw a point where $x = 4$ and $y = 3$.

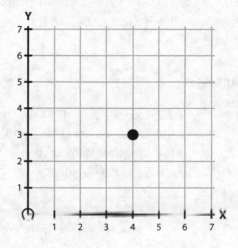

Sometimes after plotting the point students may be asked to label the points with the coordinates of the point $(4, 3)$, or they may be asked to label the point with a single letter, such as A. It is important to realize that every point on the plane is identified with exactly one unique, ordered pair.

ALERT

A common mistake that students of all ages make is to reverse the order of the coordinates when plotting points or reading coordinates from a coordinate plane. The point (7, 0) is on the x-axis, and the point (0, 7) is on the y-axis.

Graphing Points

Students record data on a coordinate plane and gain understanding for using a graph. By creating graphs where they identify the x-axis and y-axis and provide a good title, students gain skills for interpreting graphs that they may encounter in newspapers, magazines, or online. They gain understanding that the units measuring the x-axis are usually different than those of the y-axis. With experience, they observe that if time is on one of the axes, it is usually on the x-axis, and if money is a label for one of the axes it is usually on the y-axis.

EXAMPLE 29

The following graph represents the average temperate for St. Louis. The top line represents the average high temperature, and the bottom line represents the average low temperature. What is the average high temperature in St. Louis in the month of June? All temperatures shown are in Fahrenheit.

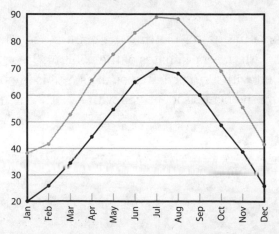

Discussion

The question asks for the high temperature for June. Look along the x-axis until you get to June. The question asks for the high temperature in June, so read straight up the y-axis until the top line is reached, then use the scale along the y-axis to find the temperature. This is under the 90-degree line, and above the 80-degree line, so an answer of 85 degrees is very reasonable. If you look closely at the high temperature for August, you will notice that it is also between 80 and 90 degrees, but slightly higher than June's average temperature. Sometimes, when exact measurements can't be read directly from the axes, neighboring data points may be used to help increase the precision of an answer.

Helping Your Fifth Grader Succeed

If you can give your child experience calculating money problems using all four operations, it will help him during the fifth grade and also prepare him for sixth grade. Helping your child relate fractions to his everyday life and connect real-world graphs you see online or in print to what he is doing in school will help him to see that the math he experiences in school is valuable.

Here are some ideas for helping your child with fifth-grade math, as well as preparing him for sixth grade and middle school. Middle school introduces many new learning factors, perhaps including a new school and a new daily schedule format. You can help him with the long-term process to build the skills necessary to become a more independent learner, and build skills for self-advocacy. Start talking to him about what it means to take responsibility for his own learning.

Around the House

As often as you can, point out to your child that math is everywhere around us. The trick to doing this is finding enjoyable ways to integrate math into everyday life without making your child feel like you are giving him more assignments.

Create graphs and charts of things that your child enjoys. If he is interested in weather, have him record the temperature or snowfall over a defined period of time. If he is interested in sports, there is a variety of books designed with math activities for most sports. If he is interested in money, have him track how much the family spends on an item, such as groceries or fast food.

Use newspapers or magazines to show your child tables, charts, or graphs that might be of interest to him. Sometimes even when the data in the graph is not interesting to a fifth grader, the *way* in which it is presented might be. (*USA Today* is a great source of interesting charts and graphs.)

Fifth graders know a lot of math! Have your child help a sibling or younger friend with math, maybe on a regular basis. Remember that students learn from what they hear, what they read, and what they say. When they discuss or explain math to another student, it deepens their understanding as well.

IMPROMPTU MATH GAMES

- When you're waiting in lines, at red lights in your car, or during commercials on TV, ask your child some multiplication facts, or ask him to tell you ten more and ten less than any number you give him up to 1,000,000. Ask him to multiply numbers by 10 or 100. This will build up his mental math facts.
- Show your child she is capable of using mental math to solve really hard problems by giving her a series of questions in a pattern. For example, start with 79 times 81. She probably can't do that off the top of her head . . . or can she? Ask her 5×5, then 4×6. Ask her 7×7, then 6×8. Ask her 10×10, then 9×11. Finally, ask her 80×80, then ask her 79×81. Does she see a pattern?

- Talk about time in fractional hours: a quarter past four, $\frac{3}{12}$ past four, three quarters until five.
- Pick a decimal number and have your child round to the nearest whole number. Give him two numbers and ask him which one is closer to a third. For example, "Is 87.50 closer to 80, or 90?"

Homework Help

Help your child manage his homework to avoid missing assignments. Some fifth graders have excellent organizational skills and time management skills, while others are less well organized and see homework as a nuisance that interferers with video games and texting. Help your child keep his notebooks/binders/backpacks organized and help him track all assignments; when were they assigned, what was assigned, when are they due?

ESSENTIAL

Many students don't know how to prepare for a math test. Teachers may provide practice problems, and those should be reviewed with your child. Reviewing notes may be helpful, but the best preparation may be to complete additional practice problems where the answers can be checked.

As part of a periodic homework check, have your child explain how he got one or more of his answers, and stress the importance of being able to explain his work to others. Help your child with math vocabulary. Ask him to define numerator and denominator, tenths and hundredths place, x-axis and y-axis. Show him a variety of shapes and ask him to classify them.

Standardized Test Preparation

Neither teachers nor parents like the idea of teaching to a test, but the Common Core Standards have identified math topics that are important for students to know for being future members of the workforce and for being prepared for college. The standards identify skills that students should know, and the standardized testing measures how proficient your child has become within the major domains of the standards. As new testing procedures are rolled out, it is important that your child be comfortable in how the tests are administered. Here are some tips for preparing for standardized tests.

- Both you and your child can become familiar with taking online tests by doing practice problems supplied by either PARCC or Smarter Balance, regardless of whether your state uses either of these or uses a test of their own. You will see differences in how the electronic assessments measure the *process* of solving a problem, which is well beyond selecting the correct answer. Students must be proficient in using the tools and techniques for taking the test.

- Math websites such as *IXL.com* can provide practice problems measuring hundreds of individual skills targeted to specific grades, and correlated to state and Common Core standards. They can provide practice and provide feedback for learning each skill.
- Use skills checklists provided by your child's teacher or found online that identify the critical skills your child should have.
- Analyze the results returned to you about your child's progress. Look at each of the domains for your child's assessment, and identify areas of strength and any areas where she needs improvement. Come up with a game plan to improve skills where needed. Let your child know where she is strong so she can build confidence.

Exercises

1. Chad and Megan were working on the following homework problem: Evaluate the expression $11 - 1 \times 5 - 4$. Chad said the answer is 10, and Megan said the answer is 46. Who do you agree with? Explain your answer.

2. What is the value of the following expression?

   $(50 + 90) \div 10$

   A. 5   C. 59

   B. 14   D. 150

3. What is the value of the following expression?

   $52 - (9 \times 5)$

   A. 7   C. 38

   B. 14   D. 215

4. Marie evaluated an expression by subtracting 9 from 17 and then multiplying the result by 4. Which of the following could be the expression Marie evaluated?

   A. $17 \times 4 - 9$ C. $4 \times (17 - 9)$

   B. $(9 + 17) \times 4$ D. $9 \times (17 - 4)$

5. Match each expression in the left column to an equivalent expression in the right column.

$$1{,}000 \qquad 10^2$$
$$100{,}000 \qquad 10^3$$
$$1{,}000{,}000 \qquad 10^4$$
$$100 \qquad 10^5$$
$$10{,}000 \qquad 10^6$$

6. Round each of the following numbers as directed.

   A. Round 8.848 to the nearest 10.

   B. Round 8.848 to the nearest tenth.

   C. Round 88.428 to the nearest whole number.

   D. Round 88.458 to the hundredth.

7. Check all of the numbers that are multiples of 7.

   ☐ 287
   ☐ 400
   ☐ 308
   ☐ 456
   ☐ 421

8. Which of the answers is equivalent to the following expression?

   $(3 \times 10) + (8 \times 1) + (5 \times 0.1) + (6 \times 0.001)$

   A. 3.856   C. 38.56

   B. 3.8506   D. 38.506

9. Jeremy measures an index card that is $6\frac{1}{2}$ cm wide and 9 cm long.

   Find the perimeter and area of Jeremy's index card.

10. Put the following fractions in order from least to greatest.

$$1\frac{1}{2} \quad \frac{3}{5} \quad \frac{5}{3} \quad 2\frac{1}{2} \quad \frac{13}{5}$$

11. The Math Club bought 2 gallons of lemonade to sell at a fundraiser to buy new math games. They sold 25 glasses, and each glass contained $5\frac{1}{2}$ ounces of lemonade. How much lemonade did they have left? (There are 32 ounces in a quart, and 4 quarts in a gallon.)

12. The Boys and Girls club has 45 ounces of ice cream. They will serve $2\frac{1}{4}$ ounces with each ice cream cone they hand out. How many ice cream cones can they hand out?

13. Walgreens is selling 8-packs of water. Each bottle has 500 milliliters of water. What is the total amount of water, in liters, of one 8-pack?

14. Howard packed the following items in a bag at the grocery store.

Item	Weight
pancake mix	1 pound 10 ounces
green beans	1 pound 7 ounces
milk	4 pounds 1 ounce

1 pound = 16 ounces

What is the total weight of the items Howard packed in the bag?

15. A large fish tank was recently installed in the public library. The width of the tank is 4 feet, the length is 9 feet, and the height is 2 feet. Find the volume of the fish tank.

16. A large fish tank was recently installed in a shopping mall. The width of the fish tank is 4 feet, and the length is 7 feet. If the volume of the fish tank is 84 cubic feet, what is the height of the fish tank?

17. The Tangram Exercise, Part 1

On a 16×16 grid using graph paper or other materials that are available, perform the following steps.

A. Plot the points (0, 8), (8, 16), (0, 0), (16, 16).

B. Draw a line from point (0, 8) to point (8, 16).

C. Draw a line from point (0, 0) to point (16, 16).

    D. Draw a line from point $(0, 0)$ to point $(0, 8)$.

    E. Draw borders around your grid: Four lines from $(0, 0)$ to $(0, 16)$, $(0, 16)$ to $(16, 16)$, $(16, 16)$ to $(16, 0)$, and $(0, 0)$ to $(16, 0)$.

18. The Tangram Exercise, Part 2

    A. Using the same grid, plot the points $(4, 4)$, $(4, 12)$, $(12, 12)$ and $(16, 0)$.

    B. Draw a line from point $(4, 4)$ to point $(4, 12)$.

    C. Draw a line from point $(4, 12)$ to point $(16, 0)$.

    D. Draw a line from point $(8, 16)$ to point $(12, 12)$.

19. The Tangram Exercise, Part 3

    A. Cut out the tangram puzzle pieces from Part 2 (there will be seven).

    B. Using exactly three pieces, see if you can make a square.

    C. Using exactly three pieces, see if you can make a rectangle.

    D. Using exactly three pieces, see if you can make a trapezoid.

    E. Which of the previous three shapes can you make using exactly two pieces?

20. Which of the following types of quadrilateral *always* has perpendicular sides?

    A. rectangle   C. parallelogram

    B. rhombus   D. trapezoid

CHAPTER 10

Helping Your Child to Succeed

Each chapter has provided tips for how to help you to help your child succeed. Though some tips may be appropriate for multiple grades, the tips that connect to multiple grades—maybe even all grades—have been left to this closing chapter. Here you'll find tips not just for helping your child succeed in math, but succeed in school. The most important academic skill you can help your child develop is reading! If you can read with your child every night, you will help him to succeed in every subject.

Communication

One of the most important things you can do is to work to keep the two-way communication lines always open with your child, establish a connection with your child's teacher and school, and, when necessary, advocate for your child.

Communicating with Your Child

Always be enthusiastic about math! If you have a daughter, make sure she knows that girls are just as smart as boys, and they can be good at math, too! If you have a son, make sure he knows that boys are just as smart as girls, and they can be good at math, too! Be generous with praise and give positive feedback; it is a great confidence builder.

Point out times throughout the day when you apply math, for example:

- "We are leaving in five minutes."
- "The chicken will be done cooking in twenty-five minutes; we can eat dinner five minutes after that."
- "If we put $20.00 of gas in the car, approximately how many gallons would we get?"
- Even performing a multistep process can be connected to math. What is the sequence for taking a bath? Put the steps in order; there is an algorithm, a set sequence of instructions to achieve a certain goal. Life is all about algorithms. Kids appreciate routines; you could map out their routine for a school day with them as a process for completing a day.

FACT

Don't say, "I'm not good at math" or "Some people are just good at math." Math is a skill like any other: It takes practice and perseverance to learn. Some people may learn math more easily that others, but not everyone's learning style is the same.

Reading with Your Child

A well-cited study identified that by the age of three, children born into low-income families heard roughly 30 million fewer words than their more affluent peers. Recent updates and subsequent studies have found that the 30-million word gap correlates to achievement gaps, and has long term consequences. Reading to or with your child helps to develop reading and listening skills, builds vocabulary, and encourages open communication channels.

Homework Help

At some point a child has to take responsibility for their own learning, but that time has not come yet in your child's development. Children need encouragement and guidance as they build their learning skills. Encourage your child to always do her best. All assignments should be completed as neatly as possible, and represent her best effort. Careless errors can often be avoided by neat work, and as she learns to check her work she will find tidy work easier to follow. There are important math practices that rely on work that is neat—it is more precise and easier to follow, therefore it improves and encourages communication.

Keep in mind the third Standard for Mathematical Practice: *Construct viable arguments and critique the reasoning of others*. Other people must be able to follow the work and be convinced of its accuracy based solely on what is presented. This also ties to the first Standard for Mathematical Practice: *Make sense of problems and persevere in solving them*. Students will be in a better position to monitor their work and progress if they have it well organized.

Help with the math content. Whenever possible, review or guide your child through the answers by asking questions, and answering her questions with questions to help her uncover the things she knows. Here parents need to persevere as well as children. Help your child find a starting point, or an entry into the problem. Ask her questions about what information is provided in the problem. Ask her about the things that you can figure out from the information in the problem (even if the problem doesn't ask you to). You will reach a point where you both will understand what the problem is asking her to do. If you reach a point where *you* don't understand what the problem is asking your child to do, then refer back to the math program resources for assistance. If you do an Internet search specific to the question, you will likely find an explanation about what the problem is asking. You can also post your question on online tutoring bulletin boards.

ESSENTIAL

You are an important resource for your child! Many parents will be hesitant to help their child, not because pre-K to grade 5 math is difficult, but because they feel the way they were taught math is too different from the way the teacher is presenting math.

Check in with your child to keep assignments organized. Know the due dates for long-term projects; help your child set milestones to help keep long-term assignments on schedule, and help your child plan time to work on them along with her other time commitments.

Help to keep your child organized. This includes notebooks, daily notes, backpacks, agenda books, school supplies at home, and anything else that causes your child to use time inefficiently. (I can't add up all the time I've lost looking for my sons' shoes; if they'd leave them in the same place each day, it would save us all valuable time.)

Coping

Does your child come home stressed from school? Are there other anxiety issues? Are there stress points in your child's life beyond the school walls? Does your child have healthy eating habits? We all want our children to grow up happy, and healthy. Often the circumstances outside of the classroom impact the success of children inside the school walls, and most often, beyond your child's control. If necessary, seek stress reduction strategies.

Communicating with Your Child's Teacher(s)

Get to know your child's math teacher. Maybe it will be the regular classroom teacher, or perhaps it will be a teacher dedicated to teaching math. An additional math coach may also be a resource for you at the school or district level. A back-to-school information night is usually held early in the school year and will provide you with the opportunity to get contact information, along with additional resources such as specifics about the math program, websites associated with the math class, and online math resources. Send an e-mail to your child's math teacher so he/she has an electronic record of your e-mail address, which makes it easier for the teacher to contact you electronically.

ESSENTIAL

E-mail is the method preferred by most teachers for communication. Addressing a concern via e-mail gives more time for the teacher to prepare a response than an impromptu phone call. When a phone call is necessary, set up a convenient time for both of you by sending an e-mail containing your availability.

A back-to-school information night is not a good venue for a one-on-one conversation with a teacher, so send the teacher an e-mail explaining that you would like to help your child at home, and that you would like a list of any parent resources that are available with your child's math program. Most math programs provide online resources or include letters to parents at the start of a new unit that will contain tips about what the parents can do to help. Make sure that you have online access to the resources that are connected to the math program, as well as links to any class websites that the teacher maintains for posting homework, grades, or important notices.

Accessing Data

Use the teacher as a resource for monitoring the progress of your student. Often the information provided by a report card or periodic interim report of progress doesn't provide the feedback within a time period that would be most effective. Sometimes a positive update does not indicate deficiencies, and a negative report does not indicate strengths. Try to be aware of both your child's strengths and those specific areas that need improvement. One of your goals should be turning your child's weaknesses to strengths while making sure he retains his pre-existing skills.

Communicating with Your Child's School

Your child's school is a very busy place. Stay informed by reading newsletters and other correspondence that is sent home with your child. Parent support is needed for a variety of extra programs that schools are able to provide. Parent-teacher organizations provide a means for meeting other parents and keeping up to date with the latest school activity.

ESSENTIAL

Find out whether your child will experience standardized testing during the school year, and stay abreast of the test dates. Practicing test questions in the form that they will be given (paper or electronic) can help reduce your child's anxiety, if there is any.

For math-specific concerns, your first step should be with the teacher directly responsible for math instruction. If your child has a primary teacher in addition to their math teacher, then it may be appropriate to address your concerns with them if you feel your concerns are not being fully addressed by the math teacher. Often there are many teachers and specialists that will know your child, and if there is an issue it may not be isolated to math; a team may be able to identify and address specific concerns.

As a last resort, the school's administration will be able to assemble all of the resources available to address any concerns that you have.

Supporting Your Child at Home

As you have seen, the last section of each chapter contains ideas for using math around the house. Here are a few more ways to integrate math into daily routines.

Math Around the House

- At the beginning of a football game, ask your child to predict the final score, then ask him how much the winning team will win by. During the game, ask your child how many points each team will need to score for his prediction to come true.

- When going to the store, have your child estimate what the total cost will be for the items to be purchased. This can be a great exercise in both rounding and estimating, along with adding and multiplying. Estimation is one of the tools covered by the fifth Standard for Mathematical Practice 5: *Use appropriate tools strategically*.
- Estimate everything, measure everything. How many minutes will it take us to get to school? How many miles is the drive to the mall? How long is it to drive to the mall (in miles)? Practice using the proper tools for measurement, and teach your child to tell time using both analog and digital watches/clocks. Throw in some unit conversions that are grade appropriate: What fraction of an hour will it take us to get to school? How many feet is it from here to the restaurant?

Games

Games are a great place for your child to apply reason and strategies. Most board games involve counting of some kind, or determining who ends the game with the most or the fewest of something. Number-based games that do not necessarily require mathematical processes are equally as important as those that do, because they require players to think about numbers outside of equations and problems. They utilize problem-solving skills and logical thinking that can be applied to many aspects of daily life.

- Sudoku is a great game for children and adults, and puzzles are available in a variety of formats and skill levels. Though most games (such as the ones in newspapers) use numbers, the way to solve the puzzles is by applying logic and reasoning, not calculations.
- SET is a card game that requires players to create sets of objects based on color, the quantity of the shapes on the card, the way that the shapes are filled in (or not filled in), and the shapes on the card. It can also be played online at *www.setgame.com*.

Model Study Habits

If you are doing office work or reading a book, model good study habits by finding a quiet place to work. If you are going to the store, make a list of the things that you need—this models organization and the way you would like your child to track assignments. Show your child some of the study habits and strategies that work for you. Do you like graphic organizers? Do you like note cards, flip books, flash cards? Do you like to make lists of things?

Let your children see you read for pleasure. Children often think that adults do not have homework, so it is important that they learn that studying and learning is a lifelong goal. Seeing you take time to study and practice will help them with their own work.

Reading Math with Your Child

The importance of reading cannot be overstated. Sure, there are children's books that use math and counting, but it is critical for your child to have at least appropriate grade-level reading skills. Standardized tests as well as school-based assessments will begin to be less calculation-based and more language-based. This will be true for online as well as paper-based tests. Your child will need to read situational problems critically, be able to extract the information given, and decode what the question is asking her to achieve. Your child will also be expected to write complete explanations to back up her answers as the standardized tests move towards performance-based assessments.

Giving students problems that require interpretation gives them the opportunity to show their thought processes and explain exactly how they are coming up with a specific answer. It is vital that your child knows how to mentally dissect everything she reads so she fully understands it. If she is unsuccessful in trying to solve a word problem, try to find out how she is understanding it. In cases of misunderstanding, you should propose different ways of thinking about or approaching a problem without giving them the answer.

Establish a Routine

Children need healthy routines—times to go to bed, times to get up, and afterschool routines. There isn't a one-size-fits all afterschool routine. Some kids like to come home from school, finish their homework, and be done with school for the rest of the day. Some students require recharging after school and need a break before doing homework or anything academic.

Many children will go to afterschool programs such as sports, academic support, academic clubs, or afterschool care. At early ages children may be required to provide support services to their family, such as staying with a younger sibling in afterschool programs, and may not be provided with an opportunity to allocate time for further academics. You can support your child by providing a routine, providing an opportunity for your child to be freed of family responsibilities, and also an opportunity to play. Keeping up with school requirements should be prioritized along with the other responsibilities.

Set both minimum and maximum limits for homework, allocating an age- and grade- appropriate amount of time. Your child's teacher did not assign four hours of homework tonight; if it is taking an unusually long amount of time to complete a task, investigate why. Some children cannot stop working at a task until it is absolutely perfect. Monitor your child's work when needed and assure them when it is "perfect" enough.

Routines, once established, are not only comfortable for children, but also give them the opportunity to find out what works best for them. Make sure your child knows that you value their routine as much as they do. Good habits, both academic and non-academic, take time to establish and not much time to disestablish.

Home and School Supplies

Non-electronic resources are just as valuable as the computer, Internet, calculators, and such things. However, it is also important that children know how to use dictionaries, thesauruses, and encyclopedias. That way, when they do not have access to a computer or calculator, they are familiar with other ways to acquire information.

Some basic school supplies, a dictionary, and maybe a calculator (if appropriate) should be available to your child. Physical dictionaries have some advantages over online dictionaries, such as reading practice, practice using alphabetized lists, indexes, and problem-solving strategies to find the correct word. Skills using a dictionary scale well to using other references, both online and offline.

Math items mentioned throughout the book include a ruler, yard stick or meter stick or both, colored pencils, index cards, markers, and pencils; graph paper, scrap paper, tracing paper; manipulatives or counters such as pennies, buttons, and like objects to count; material a bit thicker than paper, such as foam board or balsa wood that can be cut into shapes.

Environment

Children need a quiet place to work, and that can sometimes be difficult to manage. As much as we'd like to think our children are growing up in a multitasking, multimedia environment and capable of processing schoolwork and watching TV and monitoring social media via cell phones, the reality is that children are becoming more capable of doing more tasks poorly simultaneously. Children are not putting their best homework effort forward thirty seconds at a time between text messages. You will be doing yourself and your child a great service by establishing early in his education a quiet place for him to study, free of electronic games, online distractions, TV, and cell phones. Establishing such policies early will save you uphill battles later.

Organization and Time Management

Organization is important for your child to be successful. Just as working neatly on paper helps her show her work and thought process most effectively, it is important for her to have an organized work space. Without the distraction of constantly trying to make space for her books and homework, your child will be able to focus more of her attention on her work rather than what is going on around her, or being distracted by something unrelated that is sitting on the table.

Similarly, you should, as much as possible, help your child learn how to best manage her time and use it wisely. Would she rather complete the simpler tasks first so she feels like she is getting work done? Or would she rather do the harder assignments first to get them out of the way? Different methods work for different students and it is, ultimately, up to you and your child to discover what works best for her; it is important that she learns what that is.

Time-management skills are important in every aspect of life, and your child is never too young to start learning how to prioritize. When putting the groceries away, is it more important to put the milk or crackers away first? Should she start her homework by finishing her daily assignments, or should she devote most of her time to long term projects? In time, she will learn how to balance both.

Summary

Indeed, the way you were taught math will not be the same way that your child will *experience* math. Maybe most of your teachers used a direct teaching method: You sat at a desk, listened, watched, and took notes while your teacher showed you how to solve a problem, and then everyone repeated a similar problem with different numbers by following the same algorithm. Some math department heads are still insisting that their teachers stick to the tried-and-true direct teaching method, but these department heads are a retiring or evolving breed. Your child will be provided with a more discovery-based, exploratory, group-oriented method of learning that will include technology. The specific content standards will not be substantially different from what you were taught; however, the sequence in which topics are taught and the emphasis at each grade level will be different. The biggest difference will be incorporating the Standards for Mathematical Practice into action to help your child gain a richer understanding, appreciation—and hopefully, enjoyment—of math.

Common Core State Standards for Mathematics

Prekindergarten

COUNTING AND CARDINALITY

Know number names and the counting sequence.

1. Listen to and say the names of numbers in meaningful contexts.
2. Recognize and name written numerals 0–10.

Count to tell the number of objects.
Understand the relationships between numerals and quantities up to ten.

Compare numbers.

3. Count to answer "how many" questions about as many as ten items arranged in a line, a rectangular array, or a circle. Count (up to 5) items that are scattered (individual states vary among counting 5, 7, or 10 items).
4. Use comparative language, such as more/less than, equal to, to compare and describe two collections of objects.
5. Use "first" or "last" related to order or position.

OPERATIONS AND ALGEBRAIC THINKING

Understand addition as putting together and adding to, and understand subtraction as taking apart and taking from.

1. Use concrete objects to model real-world addition (putting together) and subtraction (taking away) problems up through five.

MEASUREMENT AND DATA

Describe and compare measurable attributes.

1. Recognize the attributes of length, area, weight, and capacity of everyday objects using appropriate vocabulary (e.g., long, short, tall, heavy, light, big, small, wide, narrow).
2. Compare the attributes of length and weight for two objects, including longer/shorter, same length; heavier/lighter, same weight; holds more/less, holds the same amount.
3. Classify objects and count the number of objects in each category.
4. Sort, categorize, and classify objects by more than one attribute.

5. Work with money. Recognize that certain objects are coins and that dollars and coins represent money.

GEOMETRY
Identify and describe shapes (squares, circles, triangles, rectangles).

1. Describe objects in the environment using names of shapes. Identify relative positions of objects in space using appropriate language (e.g., inside, outside, above, below, next to, behind, in front of, over, under).

2. Correctly name shapes, regardless of size.

Analyze, compare, create, and compose shapes.

3. Analyze, compare, sort two- and three-dimensional shapes and objects in different ways, using informal language to describe similarities, differences, and other attributes (e.g., by color, shape, size).

4. Create and build shapes from components, represent three-dimensional shapes (ball/sphere, square box/cube, tube/cylinder) using various manipulative materials (such as popsicle sticks, blocks, pipe cleaners, pattern blocks).

Kindergarten

COUNTING AND CARDINALITY
Know number names and the count sequence.

1. Count to 100 by ones and by tens.

2. Count forward beginning from a given number within the known sequence (instead of having to begin at 1).

3. Write numbers from 0 to 20. Represent a number of objects with a written numeral 0–20 (with 0 representing a count of no objects).

Count to tell the number of objects.

4. Understand the relationship between numbers and quantities; connect counting to cardinality.

    A. When counting objects, say the number names in the standard order, pairing each object with one and only one number name and each number name with one and only one object.

    B. Understand that the last number name said tells the number of objects counted. The number of objects is the same regardless of their arrangement or the order in which they were counted.

    C. Understand that each successive number name refers to a quantity that is one larger.

5. Count to answer "how many?" questions about as many as 20 things arranged in a line, a rectangular array, or a circle, or as many as 10 things in a scattered configuration; given a number from 1–20, count out that many objects.

Compare numbers.

6. Identify whether the number of objects in one group is greater than, less than, or equal to the number of objects in another group, e.g., by using matching and counting strategies.

7. Compare two numbers between 1 and 10 presented as written numerals.

OPERATIONS AND ALGEBRAIC THINKING
Understand addition, and understand subtraction.

1. Represent addition and subtraction with objects, fingers, mental images, drawings, sounds (e.g., claps), acting out situations, verbal explanations, expressions, or equations.

2. Solve addition and subtraction word problems, and add and subtract within 10, e.g., by using objects or drawings to represent the problem.

3. Decompose numbers less than or equal to 10 into pairs in more than one way, e.g., by using objects or drawings, and record each decomposition by a drawing or equation (e.g., $5 = 2 + 3$ and $5 = 4 + 1$).

4. For any number from 1 to 9, find the number that makes 10 when added to the given number, e.g., by using objects or drawings, and record the answer with a drawing or equation.

5. Fluently add and subtract within 5.

NUMBER AND OPERATIONS IN BASE TEN
Work with numbers 11–19 to gain foundations for place value.

1. Compose and decompose numbers from 11 to 19 into ten ones and some further ones, e.g., by using objects or drawings, and record each composition or decomposition by a drawing or equation (such as $18 = 10 + 8$); understand that these numbers are composed of ten ones and one, two, three, four, five, six, seven, eight, or nine ones.

MEASUREMENT AND DATA
Describe and compare measurable attributes.

1. Describe measurable attributes of objects, such as length or weight. Describe several measurable attributes of a single object.

2. Directly compare two objects with a measurable attribute in common, to see which object has "more of"/"less of" the attribute, and describe the difference. *For example, directly compare the heights of two children and describe one child as taller/shorter.*

Classify objects and count the number of objects in each category.

3. Classify objects into given categories; count the numbers of objects in each category and sort the categories by count.

GEOMETRY
Identify and describe shapes.

1. Describe objects in the environment using names of shapes, and describe the relative positions of these objects using terms such as *above*, *below*, *beside*, *in front of*, *behind*, and *next to*.

2. Correctly name shapes regardless of their orientations or overall size.

3. Identify shapes as two-dimensional (lying in a plane, "flat") or three-dimensional ("solid").

Analyze, compare, create, and compose shapes.

4. Analyze and compare two- and three-dimensional shapes, in different sizes and orientations, using informal language to describe their similarities, differences, parts (e.g., number of sides and vertices/"corners") and other attributes (e.g., having sides of equal length).

5. Model shapes in the world by building shapes from components (e.g., sticks and clay balls) and drawing shapes.

6. Compose simple shapes to form larger shapes. *For example, "Can you join these two triangles with full sides touching to make a rectangle?"*

Grade 1

OPERATIONS AND ALGEBRAIC THINKING
Represent and solve problems involving addition and subtraction.

1. Use addition and subtraction within 20 to solve word problems involving situations of adding to, taking from, putting together, taking apart, and comparing, with unknowns in all positions, e.g., by using objects, drawings, and equations with a symbol for the unknown number to represent the problem.

2. Solve word problems that call for addition of three whole numbers whose sum is less than or equal to 20, e.g., by using objects, drawings, and equations with a symbol for the unknown number to represent the problem.

Understand and apply properties of operations and the relationship between addition and subtraction.

3. Apply properties of operations as strategies to add and subtract. *Examples: If $8 + 3 = 11$ is known, then $3 + 8 = 11$ is also known. (Commutative property of addition.) To add $2 + 6 + 4$, the second two numbers can be added to make a ten, so $2 + 6 + 4 = 2 + 10 = 12$. (Associative property of addition.)*

4. Understand subtraction as an unknown-addend problem. *For example, subtract $10 - 8$ by finding the number that makes 10 when added to 8.*

Add and subtract within 20.

5. Relate counting to addition and subtraction (e.g., by counting on 2 to add 2).

6. Add and subtract within 20, demonstrating fluency for addition and subtraction within 10. Use strategies such as counting on; making ten (e.g., $8 + 6 = 8 + 2 + 4 = 10 + 4 = 14$); decomposing a number leading to a ten (e.g., $13 - 4 = 13 - 3 - 1 = 10 - 1 = 9$); using the relationship between addition and subtraction (e.g., knowing that $8 + 4 = 12$, one knows $12 - 8 = 4$); and creating equivalent but easier or known sums (e.g., adding $6 + 7$ by creating the known equivalent $6 + 6 + 1 = 12 + 1 = 13$).

Work with addition and subtraction equations.

7. Understand the meaning of the equal sign, and determine if equations involving addition and subtraction are true or false. For example, which of the following equations are true and which are false? $6 = 6$, $7 = 8 - 1$, $5 + 2 = 2 + 5$, $4 + 1 = 5 + 2$.

8. Determine the unknown whole number in an addition or subtraction equation relating three whole numbers. For example, determine the

unknown number that makes the equation true in each of the equations $8 + ? = 11$, $5 = \underline{\quad} - 3$, $6 + 6 = \underline{\quad}$.

NUMBER AND OPERATIONS IN BASE TEN

Extend the counting sequence.

1. Count to 120, starting at any number less than 120. In this range, read and write numerals and represent a number of objects with a written numeral.

Understand place value.

2. Understand that the two digits of a two-digit number represent amounts of tens and ones. Understand the following as special cases:

    A. 10 can be thought of as a bundle of ten ones—called a "ten."

    B. The numbers from 11 to 19 are composed of a ten and one, two, three, four, five, six, seven, eight, or nine ones.

    C. The numbers 10, 20, 30, 40, 50, 60, 70, 80, 90 refer to one, two, three, four, five, six, seven, eight, or nine tens (and 0 ones).

3. Compare two two-digit numbers based on meanings of the tens and ones digits, recording the results of comparisons with the symbols >, =, and <.

Use place value understanding and properties of operations to add and subtract.

4. Add within 100, including adding a two-digit number and a one-digit number, and adding a two-digit number and a multiple of 10, using concrete models or drawings and strategies based on place value, properties of operations, and/or the relationship between addition and subtraction; relate the strategy to a written method and explain the reasoning used. Understand that in adding two-digit numbers, one adds tens and tens, ones and ones; and sometimes it is necessary to compose a ten.

5. Given a two-digit number, mentally find 10 more or 10 less than the number, without having to count; explain the reasoning used.

6. Subtract multiples of 10 in the range 10–90 from multiples of 10 in the range 10–90 (positive or zero differences), using concrete models or drawings and strategies based on place value, properties of operations, and/or the relationship between addition and subtraction; relate the strategy to a written method and explain the reasoning used.

MEASUREMENT AND DATA

1. Measure lengths indirectly and by iterating length units. Order three objects by length; compare the lengths of two objects indirectly by using a third object.

2. Express the length of an object as a whole number of length units, by laying multiple copies of a shorter object (the length unit) end to end; understand that the length measurement of an object is the number of same-sized length units that span it with no gaps or overlaps. *Limit to contexts where the object being measured is spanned by a whole number of length units with no gaps or overlaps.*

Tell and write time.

3. Tell and write time in hours and half-hours using analog and digital clocks.

Represent and interpret data.

4. Organize, represent, and interpret data with up to three categories; ask and answer questions about the total number of data points, how many in each category, and how many more or less are in one category than in another.

GEOMETRY
Reason with shapes and their attributes.

1. Distinguish between defining attributes (e.g., triangles are closed and three-sided) versus non-defining attributes (e.g., color, orientation, overall size); build and draw shapes to possess defining attributes.
2. Compose two-dimensional shapes (rectangles, squares, trapezoids, triangles, half-circles, and quarter-circles) or three-dimensional shapes (cubes, right rectangular prisms, right circular cones, and right circular cylinders) to create a composite shape, and compose new shapes from the composite shape.
3. Partition circles and rectangles into two and four equal shares, describe the shares using the words *halves*, *fourths*, and *quarters*, and use the phrases *half of*, *fourth of*, and *quarter of*. Describe the whole as two of, or four of the shares. Understand for these examples that decomposing into more equal shares creates smaller shares.

Grade 2

OPERATIONS AND ALGEBRAIC THINKING
Represent and solve problems involving addition and subtraction.

1. Use addition and subtraction within 100 to solve one- and two-step word problems involving situations of adding to, taking from, putting together, taking apart, and comparing, with unknowns in all positions, e.g., by using drawings and equations with a symbol for the unknown number to represent the problem

Add and subtract within 20.

2. Fluently add and subtract within 20 using mental strategies. By end of Grade 2, know from memory all sums of two one-digit numbers.

Work with equal groups of objects to gain foundations for multiplication.

3. Determine whether a group of objects (up to 20) has an odd or even number of members, e.g., by pairing objects or counting them by 2s; write an equation to express an even number as a sum of two equal addends.
4. Use addition to find the total number of objects arranged in rectangular arrays with up to 5 rows and up to 5 columns; write an equation to express the total as a sum of equal addends.

NUMBER AND OPERATIONS IN BASE TEN
Understand place value.

1. Understand that the three digits of a three-digit number represent amounts of hundreds, tens, and ones; e.g., 706 equals 7 hundreds, 0 tens, and 6 ones. Understand the following as special cases:

   A. 100 can be thought of as a bundle of ten tens—called a "hundred."

   B. The numbers 100, 200, 300, 400, 500, 600, 700, 800, 900 refer to one, two, three, four, five, six, seven, eight, or nine hundreds (and 0 tens and 0 ones).

2. Count within 1000; skip-count by 5s, 10s, and 100s.
3. Read and write numbers to 1000 using base-ten numerals, number names, and expanded form.
4. Compare two three-digit numbers based on meanings of the hundreds, tens, and ones digits, using >, =, and < symbols to record the results of comparisons.

Use place value understanding and properties of operations to add and subtract.

5. Fluently add and subtract within 100 using strategies based on place value, properties of operations, and/or the relationship between addition and subtraction.
6. Add up to four two-digit numbers using strategies based on place value and properties of operations.
7. Add and subtract within 1000, using concrete models or drawings and strategies based on place value, properties of operations, and/or the relationship between addition and subtraction; relate the strategy to a written method. Understand that in adding or subtracting three-digit numbers, one adds or subtracts hundreds and hundreds, tens and tens, ones and ones; and sometimes it is necessary to compose or decompose tens or hundreds.
8. Mentally add 10 or 100 to a given number 100–900, and mentally subtract 10 or 100 from a given number 100–900.
9. Explain why addition and subtraction strategies work, using place value and the properties of operations

MEASUREMENT AND DATA
Measure and estimate lengths in standard units.

1. Measure the length of an object by selecting and using appropriate tools such as rulers, yardsticks, meter sticks, and measuring tapes.
2. Measure the length of an object twice, using length units of different lengths for the two measurements; describe how the two measurements relate to the size of the unit chosen
3. Estimate lengths using units of inches, feet, centimeters, and meters.
4. Measure to determine how much longer one object is than another, expressing the length difference in terms of a standard length unit.

Relate addition and subtraction to length.

5. Use addition and subtraction within 100 to solve word problems involving lengths that are given in the same units, e.g., by using drawings (such as drawings of rulers) and

equations with a symbol for the unknown number to represent the problem.

6. Represent whole numbers as lengths from 0 on a number line diagram with equally spaced points corresponding to the numbers $0, 1, 2, \ldots$, and represent whole-number sums and differences within 100 on a number line diagram.

Work with time and money.

7. Tell and write time from analog and digital clocks to the nearest five minutes, using A.M. and P.M.

8. Solve word problems involving dollar bills, quarters, dimes, nickels, and pennies, using $ and ¢ symbols appropriately. Example: If you have 2 dimes and 3 pennies, how many cents do you have?

Represent and interpret data.

9. Generate measurement data by measuring lengths of several objects to the nearest whole unit, or by making repeated measurements of the same object. Show the measurements by making a line plot, where the horizontal scale is marked off in whole-number units.

10. Draw a picture graph and a bar graph (with single-unit scale) to represent a data set with up to four categories. Solve simple put-together, take-apart, and compare problems using information presented in a bar graph.

GEOMETRY
Reason with shapes and their attributes.

1. Recognize and draw shapes having specified attributes, such as a given number of angles or a given number of equal faces. Identify triangles, quadrilaterals, pentagons, hexagons, and cubes.

2. Partition a rectangle into rows and columns of same-sized squares and count to find the total number of them.

3. Partition circles and rectangles into two, three, or four equal shares, describe the shares using the words *halves*, *thirds*, *half of*, *a third of*, etc., and describe the whole as two halves, three thirds, four fourths. Recognize that equal shares of identical wholes need not have the same shape.

Grade 3

OPERATIONS AND ALGEBRAIC THINKING
Represent and solve problems involving multiplication and division.

1. Interpret products of whole numbers, e.g., interpret 5×7 as the total number of objects in 5 groups of 7 objects each. *For example, describe a context in which a total number of objects can be expressed as 5×7.*

2. Interpret whole-number quotients of whole numbers, e.g., interpret $56 \div 8$ as the number of objects in each share when 56 objects are partitioned equally into 8 shares, or as a number of shares when 56 objects are partitioned into equal shares of 8 objects each.

For example, describe a context in which a number of shares or a number of groups can be expressed as 56 ÷ 8.

3. Use multiplication and division within 100 to solve word problems in situations involving equal groups, arrays, and measurement quantities, e.g., by using drawings and equations with a symbol for the unknown number to represent the problem

4. Determine the unknown whole number in a multiplication or division equation relating three whole numbers. *For example, determine the unknown number that makes the equation true in each of the equations 8 × ? = 48, 5 = ___ ÷ 3, 6 × 6 = ?*

Understand properties of multiplication and the relationship between multiplication and division.

5. Apply properties of operations as strategies to multiply and divide. *Examples: If 6 × 4 = 24 is known, then 4 × 6 = 24 is also known. (Commutative property of multiplication.) 3 × 5 × 2 can be found by 3 × 5 = 15, then 15 × 2 = 30, or by 5 × 2 = 10, then 3 × 10 = 30. (Associative property of multiplication.) Knowing that 8 × 5 = 40 and 8 × 2 = 16, one can find 8 × 7 as 8 × (5 + 2) = (8 × 5) + (8 × 2) = 40 + 16 = 56. (Distributive property.)*

6. Understand division as an unknown-factor problem. *For example, find 32 ÷ 8 by finding the number that makes 32 when multiplied by 8.*

Multiply and divide within 100.

7. Fluently multiply and divide within 100, using strategies such as the relationship between multiplication and division (e.g., knowing that 8 × 5 = 40, one knows 40 ÷ 5 = 8) or properties of operations. By the end of Grade 3, know from memory all products of two one-digit numbers.

Solve problems involving the four operations, and identify and explain patterns in arithmetic.

8. Solve two-step word problems using the four operations. Represent these problems using equations with a letter standing for the unknown quantity. Assess the reasonableness of answers using mental computation and estimation strategies including rounding.

9. Identify arithmetic patterns (including patterns in the addition table or multiplication table), and explain them using properties of operations. *For example, observe that 4 times a number is always even, and explain why 4 times a number can be decomposed into two equal addends.*

NUMBER AND OPERATIONS IN BASE TEN
Use place value understanding and properties of operations to perform multi-digit arithmetic.

1. Use place value understanding to round whole numbers to the nearest 10 or 100.

2. Fluently add and subtract within 1,000 using strategies and algorithms based on place value, properties of operations, and/or the relationship between addition and subtraction.

3. Multiply one-digit whole numbers by multiples of 10 in the range 10–90 (e.g., 9×80, 5×60) using strategies based on place value and properties of operations.

NUMBER AND OPERATIONS—FRACTIONS
Develop understanding of fractions as numbers.

1. Understand a fraction $\frac{1}{b}$ as the quantity formed by 1 part when a whole is partitioned into b equal parts; understand a fraction $\frac{a}{b}$ as the quantity formed by a parts of size $\frac{1}{b}$.

2. Understand a fraction as a number on the number line; represent fractions on a number line diagram.

    A. Represent a fraction $\frac{1}{b}$ on a number line diagram by defining the interval from 0 to 1 as the whole and partitioning it into b equal parts. Recognize that each part has size $\frac{1}{b}$ and that the endpoint of the part based at 0 locates the number $\frac{1}{b}$ on the number line.

    B. Represent a fraction $\frac{a}{b}$ on a number line diagram by marking off a lengths $\frac{1}{b}$ from 0. Recognize that the resulting interval has size $\frac{a}{b}$ and that its endpoint locates the number $\frac{a}{b}$ on the number line.

3. Explain equivalence of fractions in special cases, and compare fractions by reasoning about their size.

    A. Understand two fractions as equivalent (equal) if they are the same size, or the same point on a number line.

    B. Recognize and generate simple equivalent fractions, e.g., $\frac{1}{2} = \frac{2}{4}$, $\frac{4}{6} = \frac{2}{3}$. Explain why the fractions are equivalent, e.g., by using a visual fraction model.

    C. Express whole numbers as fractions, and recognize fractions that are equivalent to whole numbers. *Examples: Express 3 in the form $3 = \frac{3}{1}$; recognize that $\frac{6}{1} = 6$; locate $\frac{4}{4}$ and 1 at the same point of a number line diagram.*

    D. Compare two fractions with the same numerator or the same denominator by reasoning about their size. Recognize that comparisons are valid only when the two fractions refer to the same whole. Record the results of comparisons with the symbols >, =, or <, and justify the conclusions, e.g., by using a visual fraction model.

MEASUREMENT AND DATA
Solve problems involving measurement and estimation.

1. Tell and write time to the nearest minute and measure time intervals in minutes. Solve word

problems involving addition and subtraction of time intervals in minutes, e.g., by representing the problem on a number line diagram.

2. Measure and estimate liquid volumes and masses of objects using standard units of grams (g), kilograms (kg), and liters (l). Add, subtract, multiply, or divide to solve one-step word problems involving masses or volumes that are given in the same units, e.g., by using drawings (such as a beaker with a measurement scale) to represent the problem.

Represent and interpret data.

3. Draw a scaled picture graph and a scaled bar graph to represent a data set with several categories. Solve one- and two-step "how many more" and "how many less" problems using information presented in scaled bar graphs. *For example, draw a bar graph in which each square in the bar graph might represent 5 pets.*

4. Generate measurement data by measuring lengths using rulers marked with halves and fourths of an inch. Show the data by making a line plot, where the horizontal scale is marked off in appropriate units—whole numbers, halves, or quarters.

Geometric measurement: understand concepts of area and relate area to multiplication and to addition.

5. Recognize area as an attribute of plane figures and understand concepts of area measurement.

   A. A square with side length 1 unit, called "a unit square," is said to have "one square unit" of area, and can be used to measure area.

   B. A plane figure that can be covered without gaps or overlaps by n unit squares is said to have an area of n square units.

6. Measure areas by counting unit squares (square cm, square m, square in, square ft, and improvised units).

7. Relate area to the operations of multiplication and addition.

   A. Find the area of a rectangle with whole-number side lengths by tiling it, and show that the area is the same as would be found by multiplying the side lengths.

   B. Multiply side lengths to find areas of rectangles with whole-number side lengths in the context of solving real world and mathematical problems, and represent whole-number products as rectangular areas in mathematical reasoning.

   C. Use tiling to show in a concrete case that the area of a rectangle with whole-number side lengths a and $b + c$ is the sum of $a \times b$ and $a \times c$. Use area models to represent the distributive property in mathematical reasoning.

   D. Recognize area as additive. Find areas of rectilinear figures by decomposing them into non-overlapping rectangles and adding the areas of the non-overlapping parts, applying this technique to solve real world problems.

Geometric measurement: recognize perimeter.

8. Solve real-world and mathematical problems involving perimeters of polygons, including finding the perimeter given the side lengths, finding an unknown side length, and exhibiting rectangles with the same perimeter and different areas or with the same area and different perimeters.

GEOMETRY
Reason with shapes and their attributes.

1. Understand that shapes in different categories (e.g., rhombuses, rectangles, and others) may share attributes (e.g., having four sides), and that the shared attributes can define a larger category (e.g., quadrilaterals). Recognize rhombuses, rectangles, and squares as examples of quadrilaterals, and draw examples of quadrilaterals that do not belong to any of these subcategories.

2. Partition shapes into parts with equal areas. Express the area of each part as a unit fraction of the whole. *For example, partition a shape into 4 parts with equal area, and describe the area of each part as $\frac{1}{4}$ of the area of the shape.*

Grade 4

OPERATIONS AND ALGEBRAIC THINKING
Use the four operations with whole numbers to solve problems.

1. Interpret a multiplication equation as a comparison, e.g., interpret $35 = 5 \times 7$ as a statement that 35 is 5 times as many as 7 and 7 times as many as 5. Represent verbal statements of multiplicative comparisons as multiplication equations.

2. Multiply or divide to solve word problems involving multiplicative comparison, e.g., by using drawings and equations with a symbol for the unknown number to represent the problem, distinguishing multiplicative comparison from additive comparison.

3. Solve multistep word problems posed with whole numbers and having whole-number answers using the four operations, including problems in which remainders must be interpreted. Represent these problems using equations with a letter standing for the unknown quantity. Assess the reasonableness of answers using mental computation and estimation strategies including rounding.

Gain familiarity with factors and multiples.

4. Find all factor pairs for a whole number in the range 1–100. Recognize that a whole number is a multiple of each of its factors. Determine whether a given whole number in the range 1–100 is a multiple of a given one-digit number. Determine whether a given whole number in the range 1–100 is prime or composite.

Generate and analyze patterns.

5. Generate a number or shape pattern that follows a given rule. Identify apparent features of the pattern that were not explicit in the rule itself. *For example, given the rule "Add 3" and the starting number 1, generate terms in the resulting sequence and observe that the terms appear to alternate between odd and even numbers. Explain informally why the numbers will continue to alternate in this way.*

NUMBER AND OPERATIONS IN BASE TEN

Generalize place value understanding for multi-digit whole numbers.

1. Recognize that in a multi-digit whole number, a digit in one place represents ten times what it represents in the place to its right. *For example, recognize that $700 \div 70 = 10$ by applying concepts of place value and division.*

2. Read and write multi-digit whole numbers using base-ten numerals, number names, and expanded form. Compare two multi-digit numbers based on meanings of the digits in each place, using $>$, $=$, and $<$ symbols to record the results of comparisons.

3. Use place value understanding to round multi-digit whole numbers to any place.

Use place value understanding and properties of operations to perform multi-digit arithmetic.

4. Fluently add and subtract multi-digit whole numbers using the standard algorithm.

5. Multiply a whole number of up to four digits by a one-digit whole number, and multiply two two-digit numbers, using strategies based on place value and the properties of operations. Illustrate and explain the calculation by using equations, rectangular arrays, and/or area models.

6. Find whole-number quotients and remainders with up to four-digit dividends and one-digit divisors, using strategies based on place value, the properties of operations, and/or the relationship between multiplication and division. Illustrate and explain the calculation by using equations, rectangular arrays, and/or area models.

NUMBER AND OPERATIONS—FRACTIONS

Extend understanding of fraction equivalence and ordering.

1. Explain why a fraction $\frac{a}{b}$ is equivalent to a fraction $\frac{(n \times a)}{(n \times b)}$ by using visual fraction models, with attention to how the number and size of the parts differ even though the two fractions themselves are the same size. Use this principle to recognize and generate equivalent fractions.

2. Compare two fractions with different numerators and different denominators, e.g., by creating common denominators or numerators, or by comparing to a benchmark fraction such as $\frac{1}{2}$. Recognize that comparisons are valid only when the two fractions refer to the same whole. Record the results of comparisons with symbols $>$, $=$, or $<$, and justify the conclusions, e.g., by using a visual fraction model.

Build fractions from unit fractions by applying and extending previous understandings of operations on whole numbers.

3. Understand a fraction $\frac{a}{b}$ with $a > 1$ as a sum of fractions $\frac{1}{b}$.

A. Understand addition and subtraction of fractions as joining and separating parts referring to the same whole.

B. Decompose a fraction into a sum of fractions with the same denominator in more than one way, recording each decomposition by an equation. Justify decompositions, e.g., by using a visual fraction model. *Examples:* $\frac{3}{8} = \frac{1}{8} + \frac{1}{8} + \frac{1}{8}$; $\frac{3}{8} = \frac{1}{8} + \frac{2}{8}$; $2\frac{1}{8} = 1 + 1 + \frac{1}{8} = \frac{8}{8} + \frac{8}{8} + \frac{1}{8}$.

C. Add and subtract mixed numbers with like denominators, e.g., by replacing each mixed number with an equivalent fraction, and/or by using properties of operations and the relationship between addition and subtraction.

D. Solve word problems involving addition and subtraction of fractions referring to the same whole and having like denominators, e.g., by using visual fraction models and equations to represent the problem.

4. Apply and extend previous understandings of multiplication to multiply a fraction by a whole number.

A. Understand a fraction $\frac{a}{b}$ as a multiple of $\frac{1}{b}$. *For example, use a visual fraction model to represent $\frac{5}{4}$ as the product $5 \times (\frac{1}{4})$, recording the conclusion by the equation $\frac{5}{4} = 5 \times (\frac{1}{4})$.*

B. Understand a multiple of $\frac{a}{b}$ as a multiple of $\frac{1}{b}$, and use this understanding to multiply a fraction by a whole number. *For example, use a visual fraction model to express $3 \times (\frac{2}{5})$ as $6 \times (\frac{1}{5})$, recognizing this product as $\frac{6}{5}$. (In general, $n \times \frac{a}{b} = \frac{(n \times a)}{b}$.)*

C. Solve word problems involving multiplication of a fraction by a whole number, e.g., by using visual fraction models and equations to represent the problem. *For example, if each person at a party will eat $\frac{3}{8}$ of a pound of roast beef, and there will be 5 people at the party, how many pounds of roast beef will be needed? Between what two whole numbers does your answer lie?*

Understand decimal notation for fractions, and compare decimal fractions.

5. Express a fraction with denominator 10 as an equivalent fraction with denominator 100, and use this technique to add two fractions with respective denominators 10 and 100. *For example, express* $\frac{3}{10}$ *as* $\frac{30}{100}$, *and add* $\frac{3}{10}$ + $\frac{4}{100} = \frac{34}{100}$.

6. Use decimal notation for fractions with denominators 10 or 100. *For example, rewrite 0.62 as* $\frac{62}{100}$; *describe a length as 0.62 meters; locate 0.62 on a number line diagram.*

7. Compare two decimals to hundredths by reasoning about their size. Recognize that comparisons are valid only when the two decimals refer to the same whole. Record the results of comparisons with the symbols >, =, or <, and justify the conclusions, e.g., by using a visual model.

MEASUREMENT AND DATA

Solve problems involving measurement and conversion of measurements.

1. Know relative sizes of measurement units within one system of units including km, m, cm; kg, g; lb, oz.; l, ml; hr, min, sec. Within a single system of measurement, express measurements in a larger unit in terms of a smaller unit. Record measurement equivalents in a two-column table. *For example, know that 1 ft is 12 times as long as 1 in. Express the length of a 4 ft snake as 48 in.*

Generate a conversion table for feet and inches listing the number pairs (1, 12), (2, 24), (3, 36), ...

2. Use the four operations to solve word problems involving distances, intervals of time, liquid volumes, masses of objects, and money, including problems involving simple fractions or decimals, and problems that require expressing measurements given in a larger unit in terms of a smaller unit. Represent measurement quantities using diagrams such as number line diagrams that feature a measurement scale.

3. Apply the area and perimeter formulas for rectangles in real world and mathematical problems. *For example, find the width of a rectangular room given the area of the flooring and the length, by viewing the area formula as a multiplication equation with an unknown factor.*

Represent and interpret data.

4. Make a line plot to display a data set of measurements in fractions of a unit ($\frac{1}{2}$, $\frac{1}{4}$, $\frac{1}{8}$). Solve problems involving addition and subtraction of fractions by using information presented in line plots. *For example, from a line plot find and interpret the difference in length between the longest and shortest specimens in an insect collection.*

Geometric measurement: understand concepts of angles and measure angles.

5. Recognize angles as geometric shapes that are formed wherever two rays share a common

endpoint, and understand concepts of angle measurement:

A. An angle is measured with reference to a circle with its center at the common endpoint of the rays, by considering the fraction of the circular arc between the points where the two rays intersect the circle. An angle that turns through $\frac{1}{360}$ of a circle is called a "one-degree angle," and can be used to measure angles.

B. An angle that turns through n one-degree angles is said to have an angle measure of n degrees.

6. Measure angles in whole-number degrees using a protractor. Sketch angles of specified measure.

7. Recognize angle measure as additive. When an angle is decomposed into non-overlapping parts, the angle measure of the whole is the sum of the angle measures of the parts. Solve addition and subtraction problems to find unknown angles on a diagram in real-world and mathematical problems, e.g., by using an equation with a symbol for the unknown angle measure.

GEOMETRY
Draw and identify lines and angles, and classify shapes by properties of their lines and angles.

1. Draw points, lines, line segments, rays, angles (right, acute, obtuse), and perpendicular and parallel lines. Identify these in two-dimensional figures.

2. Classify two-dimensional figures based on the presence or absence of parallel or perpendicular lines, or the presence or absence of angles of a specified size. Recognize right triangles as a category, and identify right triangles.

3. Recognize a line of symmetry for a two-dimensional figure as a line across the figure such that the figure can be folded along the line into matching parts. Identify line-symmetric figures and draw lines of symmetry.

Grade 5

OPERATIONS AND ALGEBRAIC THINKING
Write and interpret numerical expressions.

1. Use parentheses, brackets, or braces in numerical expressions, and evaluate expressions with these symbols.

2. Write simple expressions that record calculations with numbers, and interpret numerical expressions without evaluating them. *For example, express the calculation "add 8 and 7, then multiply by 2" as $2 \times (8+7)$. Recognize that $3 \times (18,932+921)$ is three times as large as $18,932+921$, without having to calculate the indicated sum or product.*

Analyze patterns and relationships.

3. Generate two numerical patterns using two given rules. Identify apparent relationships between corresponding terms. Form ordered pairs consisting of corresponding terms from

the two patterns, and graph the ordered pairs on a coordinate plane. *For example, given the rule "Add 3" and the starting number 0, and given the rule "Add 6" and the starting number 0, generate terms in the resulting sequences, and observe that the terms in one sequence are twice the corresponding terms in the other sequence. Explain informally why this is so.*

NUMBER AND OPERATIONS IN BASE TEN
Understand the place value system.

1. Recognize that in a multi-digit number, a digit in one place represents 10 times as much as it represents in the place to its right and $\frac{1}{10}$ of what it represents in the place to its left.

2. Explain patterns in the number of zeros of the product when multiplying a number by powers of 10, and explain patterns in the placement of the decimal point when a decimal is multiplied or divided by a power of 10. Use whole-number exponents to denote powers of 10.

3. Read, write, and compare decimals to thousandths.

   A. Read and write decimals to thousandths using base-ten numerals, number names, and expanded form, e.g., $347.392 = 3 \times 100 + 4 \times 10 + 7 \times 1 + 3 \times (\frac{1}{10}) + 9 \times (\frac{1}{100}) + 2 \times (\frac{1}{1,000})$.

   B. Compare two decimals to thousandths based on meanings of the digits in each place, using >, =, and < symbols to record the results of comparisons.

4. Use place value understanding to round decimals to any place.

Perform operations with multi-digit whole numbers and with decimals to hundredths.

5. Fluently multiply multi-digit whole numbers using the standard algorithm.

6. Find whole-number quotients of whole numbers with up to four-digit dividends and two-digit divisors, using strategies based on place value, the properties of operations, and/or the relationship between multiplication and division. Illustrate and explain the calculation by using equations, rectangular arrays, and/or area models.

7. Add, subtract, multiply, and divide decimals to hundredths, using concrete models or drawings and strategies based on place value, properties of operations, and/or the relationship between addition and subtraction; relate the strategy to a written method and explain the reasoning used.

NUMBER AND OPERATIONS—FRACTIONS
Use equivalent fractions as a strategy to add and subtract fractions.

1. Add and subtract fractions with unlike denominators (including mixed numbers) by replacing given fractions with equivalent fractions in such a way as to produce an equivalent sum or difference of fractions with like denominators. *For example, $\frac{2}{3} + \frac{5}{4} = \frac{8}{12} + \frac{15}{12} = \frac{23}{12}$. (In general, $\frac{a}{b} + \frac{c}{d} = \frac{(ad + bc)}{bd}$.)*

2. Solve word problems involving addition and subtraction of fractions referring to the same whole, including cases of unlike denominators, e.g., by using visual fraction models or equations to represent the problem. Use benchmark fractions and number sense of fractions to estimate mentally and assess the reasonableness of answers. *For example, recognize an incorrect result* $\frac{2}{5} + \frac{1}{2} = \frac{3}{7}$, *by observing that* $\frac{3}{7} < \frac{1}{2}$.

Apply and extend previous understandings of multiplication and division.

3. Interpret a fraction as division of the numerator by the denominator ($\frac{a}{b} = a \div b$). Solve word problems involving division of whole numbers leading to answers in the form of fractions or mixed numbers, e.g., by using visual fraction models or equations to represent the problem. *For example, interpret* $\frac{3}{4}$ *as the result of dividing 3 by 4, noting that* $\frac{3}{4}$ *multiplied by 4 equals 3, and that when 3 wholes are shared equally among 4 people each person has a share of size* $\frac{3}{4}$. *If 9 people want to share a 50-pound sack of rice equally by weight, how many pounds of rice should each person get? Between what two whole numbers does your answer lie?*

4. Apply and extend previous understandings of multiplication to multiply a fraction or whole number by a fraction.

A. Interpret the product ($\frac{a}{b}$) $\times q$ as a parts of a partition of q into b equal parts; equivalently, as the result of a sequence of operations $a \times q \div b$. *For example, use a visual fraction model to show* ($\frac{2}{3}$) $\times 4 = \frac{8}{3}$, *and create a story context for this equation. Do the same with* ($\frac{2}{3}$) $\times$ ($\frac{4}{5}$) $= \frac{8}{15}$. *(In general,* ($\frac{a}{b}$) $\times$ ($\frac{c}{d}$) $= \frac{ac}{bd}$.)

B. Find the area of a rectangle with fractional side lengths by tiling it with unit squares of the appropriate unit fraction side lengths, and show that the area is the same as would be found by multiplying the side lengths. Multiply fractional side lengths to find areas of rectangles, and represent fraction products as rectangular areas.

5. Interpret multiplication as scaling (resizing), by:

A. Comparing the size of a product to the size of one factor on the basis of the size of the other factor, without performing the indicated multiplication.

B. Explaining why multiplying a given number by a fraction greater than 1 results in a product greater than the given number (recognizing multiplication by whole numbers greater than 1 as a familiar case); explaining why multiplying a given number by a fraction less than 1 results in a product smaller than the given number; and relating the principle of fraction equivalence $\dfrac{a}{b} = \dfrac{(n \times a)}{(n \times b)}$ to the effect of multiplying $\dfrac{a}{b}$ by 1.

6. Solve real-world problems involving multiplication of fractions and mixed numbers, e.g., by using visual fraction models or equations to represent the problem.

7. Apply and extend previous understandings of division to divide unit fractions by whole numbers and whole numbers by unit fractions.

   A. Interpret division of a unit fraction by a non-zero whole number, and compute such quotients. *For example, create a story context for $(\dfrac{1}{3}) \div 4$, and use a visual fraction model to show the quotient. Use the relationship between multiplication and division to explain that $(\dfrac{1}{3}) \div 4 = \dfrac{1}{12}$ because $(\dfrac{1}{12}) \times 4 = \dfrac{1}{3}$.*

   B. Interpret division of a whole number by a unit fraction, and compute such quotients. *For example, create a story context for $4 \div (\dfrac{1}{5})$, and use a visual fraction model to show the quotient. Use the relationship between multiplication and division to explain that $4 \div (\dfrac{1}{5}) = 20$ because $20 \times (\dfrac{1}{5}) = 4$.*

   C. Solve real-world problems involving division of unit fractions by non-zero whole numbers and division of whole numbers by unit fractions, e.g., by using visual fraction models and equations to represent

the problem. *For example, how much chocolate will each person get if 3 people share $\dfrac{1}{2}$ lb of chocolate equally? How many $\dfrac{1}{3}$-cup servings are in 2 cups of raisins?*

MEASUREMENT AND DATA

Convert like measurement units within a given measurement system.

1. Convert among different-sized standard measurement units within a given measurement system (e.g., convert 5 cm to 0.05 m), and use these conversions in solving multistep, real-world problems.

Represent and interpret data.

2. Make a line plot to display a data set of measurements in fractions of a unit ($\dfrac{1}{2}$, $\dfrac{1}{4}$, $\dfrac{1}{8}$). Use operations on fractions for this grade to solve problems involving information presented in line plots. *For example, given different measurements of liquid in identical beakers, find the amount of liquid each beaker would contain if the total amount in all the beakers were redistributed equally.*

Geometric measurement: understand concepts of volume.

3. Recognize volume as an attribute of solid figures and understand concepts of volume measurement.

A. A cube with side length 1 unit, called a "unit cube," is said to have "one cubic unit" of volume, and can be used to measure volume.

B. A solid figure which can be packed without gaps or overlaps using n unit cubes is said to have a volume of n cubic units.

4. Measure volumes by counting unit cubes, using cubic cm, cubic in, cubic ft, and improvised units.

5. Relate volume to the operations of multiplication and addition and solve real world and mathematical problems involving volume.

A. Find the volume of a right rectangular prism with whole-number side lengths by packing it with unit cubes, and show that the volume is the same as would be found by multiplying the edge lengths, equivalently by multiplying the height by the area of the base. Represent threefold whole-number products as volumes, e.g., to represent the associative property of multiplication.

B. Apply the formulas $V = l \times w \times h$ and $V = b \times h$ for rectangular prisms to find volumes of right rectangular prisms with whole-number edge lengths in the context of solving real world and mathematical problems.

C. Recognize volume as additive. Find volumes of solid figures composed of two non-overlapping right rectangular prisms by adding the volumes of the non-overlapping parts, applying this technique to solve real world problems.

GEOMETRY

Graph points on the coordinate plane to solve real-world and mathematical problems.

1. Use a pair of perpendicular number lines, called axes, to define a coordinate system, with the intersection of the lines (the origin) arranged to coincide with the 0 on each line and a given point in the plane located by using an ordered pair of numbers, called its coordinates. Understand that the first number indicates how far to travel from the origin in the direction of one axis, and the second number indicates how far to travel in the direction of the second axis, with the convention that the names of the two axes and the coordinates correspond (e.g., x-axis and x-coordinate, y-axis and y-coordinate).

2. Represent real world and mathematical problems by graphing points in the first quadrant of the coordinate plane, and interpret coordinate values of points in the context of the situation.

Classify two-dimensional figures into categories based on their properties.

3. Understand that attributes belonging to a category of two-dimensional figures also belong to all subcategories of that category. For example, all rectangles have four right angles and squares are rectangles, so all squares have four right angles.

4. Classify two-dimensional figures in a hierarchy based on properties.

Exercise Solutions

Chapter 6

1. The 13 frogs Tommy found can be shown by 1 group of 10 with 3 left over:

$$\text{卌卌} \quad |||$$

The 20 frogs Zach found can be shown two groups of 10:

$$\text{卌卌} \quad \text{卌卌}$$

Putting all of the frogs together gives 3 groups of 10 with 3 left over, for a total of 33 frogs.

$$\text{卌卌} \quad \text{卌卌} \quad \text{卌卌} \quad |||$$

2. The sum of this amount is $4 + 4 + 4 = 12$.

3. Using the count-up method, see that 4 would need to be added to get to 40, then 3 more is needed to count up to 43. $4 + 3 = 7$.

$$\underline{43} - \underline{36} = \underline{7}$$

4. Jazzy has 10 fewer than Helen and Helen has 60, so Jazzy has $60 - 10 = 50$.

$$\text{Jazzy} + \text{Helen} + \text{Ava} + \text{Becca} = \text{total number of hair ties}$$

$$50 + 60 + 60 + 40 = 210$$

5. Write both numbers in expanded form:

$$600 + 70 + 5 + 100 + 20 + 6$$

Reorder the numbers from hundreds, to tens, to ones:

$$600 + 100 + 70 + 20 + 5 + 6$$

Add 600 and 100 to get 700, then add 70 to get 770. Add 20 to get 790, then add 5 to get 795, then add 6 to get 801.

6. Brad has:

$$25 + 25 = 50 \text{ cents of quarters}$$
$$10 + 10 = 20 \text{ cents of dimes}$$
$$1 + 1 = 2 \text{ cents of pennies}$$
$$50 + 20 + 2 = 72 \text{ cents}$$

Chapter 7

1. 249

2. To help your child understand this problem, find 32 objects. Take 8 and spread them apart, so you model 8 groups of 1. This will leave you with 24 objects. Put one more piece in each group to model 8 groups of 2. Continue putting one object in each group until you have 8 groups of 4. You have taken a whole (the pile of 32 objects) and divided it into 8 equal groups, with 4 objects in each group. $32 \div 8 = \underline{4}$.

3. $9 \times \underline{6} = 54$

4. $121 - 78 = 43$. Gabby has 43 more stuffed animals than Jill.

5. Look at the tens place. It is a 4, so the nearest 100 will be rounded down. This gives 400. Some students may prefer to look at the last two digits (49). Since $49 < 50$, rounding to the nearest hundred will mean rounding down to 400.

6. To round to the nearest 10, look at the digit in the ones place. The digit in the ones place is a 3. Since 3 is less than 5, round down to 820.

7. Using the count-up method, Niko needs 4 to get to 280, 20 more to get to 300, and 100 more to get to 400. $4 + 20 + 100 = 124$.

8. $\frac{1}{2}$

9. Possible solutions: 48×1; 24×2; 16×3; 12×4; 8×6

10. Nine kids said their favorite team was the Spurs. Five kids said their favorite team was the Heat. $9 - 5 = 4$. Four more students said their favorite team was the Spurs rather than the Heat. The problem calls for reading the graph correctly and performing a calculation from the data read.

11. The pen has a mass of $25 + 11 = 36$ grams. 3 pencils have a total mass of $25 + 25 + 25 = 75$ grams. Two pens have a mass of $36 + 36 = 72$ grams. 3 pencils have a total mass 3 grams more than two pencils, so Jill is not correct.

12. The perimeter of the park is the sum of each side $= 140 + 140 + 51 + 51 = 382$ feet.

13. 8:05

14. Quadrilaterals have four sides. The shapes with four sides are the square, rectangle, and trapezoid.

15. A rhombus is a quadrilateral with four equal sides. The second and fifth shapes look like they have equal sides, so they are each a rhombus.

16. Bobby divided 50 pounds of sand in 5 buckets; each of the 5 buckets had 10 pounds of sand. $50 \div 5 = 10$.

17. Each rollerblade has 4 wheels, each person wears 2 rollerblades, so each person has $2 \times 4 = 8$ wheels. If Samantha and each of her 6 friends have rollerblades, then $6 + 1 = 7$ people have rollerblades. There are $8 \times 7 = 56$ wheels all together.

18. Chessboard area:

    A. There are $8 \times 8 = 64$ squares on the chessboard.

    B. The area of each square is $2 \times 2 = 4$ square inches.

    C. There are 8 squares and each square has an area of 4 square inches, so the area of each row is $8 \times 4 = 32$ square inches. Two rows have an area of $32 + 32 = 64$ square inches, 4 rows have an area of $64 + 64 = 128$ square inches, and 8 rows have an area of $128 + 128 = 256$ square inches. This is the area of the entire chessboard.

19. There are four equal parts of this garden. One part has corn, so $\frac{1}{4}$ of the area has corn.

20. Shaded fractions:

    A. 1 out of 2 parts, or $\frac{1}{2}$ of the area, is shaded.

    B. 1 out of 4 parts, or $\frac{1}{4}$ of the area, is shaded.

    C. 1 out of 3 parts, or $\frac{1}{3}$ of the area, is shaded.

Chapter 8

1. Missing numbers:

    A. $7 \times 8 = 56$

    B. $51 = 17 \times 3$

    C. 2 groups of 31 are 62.

2. Let p represent the number of posters Miss Curtis has. This is 12 more than Miss Henderson has, so Miss Henderson has 12 less, or $p - 12$. Together they have 28 posters. So $(p) + (p - 12) = 28$. Miss Curtis has 20 posters, and Miss Henderson has 12 less, which is 8. If Miss Henderson had 12 more posters they would both have the same amount, and the total would then be 40 posters.

3. Look for the missing factor. $5 \times \underline{\quad} = 55$? $5 \times 11 = 55$, so each hat cost $11.

4. Part A, multiply by 3, then subtract 1: 4, 11, 32, 95, 284, 851; Part B, multiply by 2, then add 2: 3, 8, 18, 38, 78, 158, 318, 638; Part C. Multiply by 2, then subtract 2: 5, 8, 14, 26, 50, 98, 194, 386.

5. Multiply by 10:

    A. $845 \times 10 = 8,450$

    B. $16,831 \times 10 = 168,310$

    C. $172 \times 10 = 1,720$

    D. $4,180 \times 10 = 41,800$

6. Rounded numbers:

    A. 14,800   C. 0

    B. 170   D. 146,000

7. Look at multiples of 9:

$$9 \times 10 = 90$$
$$9 \times 20 = 180$$
$$9 \times 30 = 270$$

If you were dividing 260 between 9 people, you would not have enough to give everyone

30, but you can give everyone 20. It would use up 180. $260 - 180 = 80$, so you could still give everyone some more. $9 \times 8 = 72$, so give everyone 8 more. $80 - 72 = 8$. You don't have enough to give everyone any more; you have given everyone $20 + 8 = 28$, and you have 8 left over.

8. Use an area model to show $40 + 2 \times 50 + 1$.

$$40 \times 50 = 2000$$
$$40 \times 1 = 40$$
$$2 \times 50 = 100$$
$$2 \times 1 = 2$$
$$2000 + 40 + 100 + 2 = 2{,}142$$

9. $\frac{1}{2} + 4\frac{1}{2}$ can be rewritten $\frac{1}{2} + 4 + \frac{1}{2}$ or $\frac{1}{2} + \frac{1}{2} + 4$. Since $\frac{1}{2} + \frac{1}{2} = 1$, the problem becomes $1 + 4 = 5$. Tina read 5 pages for that day.

10. Since $\frac{1}{4} + \frac{1}{4} + \frac{1}{4} + \frac{1}{4} = 1$, she needs to buy 1 pound for 4 students, 2 pounds for 8 students, and 4 pounds for 16 students. Since $24 = 16 + 8$, she bought $4 + 2 = 6$ pounds of hamburger.

11. Visual fraction model:

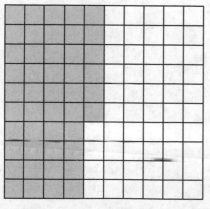

12. Fraction comparisons:

   A. $\frac{3}{4} > \frac{3}{5}$

   B. $\frac{5}{10} > \frac{5}{100}$

   C. $0.03 < 0.3$

13. Draw a rectangle, and label the top and the bottom each as 9 feet. The sum of the two lengths and the two widths has to be 108 feet. The two lengths have a sum $9 + 9 = 18$. $18 +$ the two widths has to be 108; the two widths have to add up to 90, and the two widths have to be the same. This can be viewed either as a missing sum: ___ + ___ = 90, or as a missing factors problem: $2 \times$ ___ $= 90$. Each width is 45 feet. To find the area of Peggy's room, multiply the length and the width: 45×9. On your drawing, partition the sides with 45 as $40 + 5$. Complete the area diagram to show $(40 + 5) \times 90$. $40 \times 9 = 360$; $5 \times 9 = 45$. $360 + 45 = 405$. The area of Peggy's bedroom is 405 square feet.

14. There are 12 inches in one foot, and 6 inches in $\frac{1}{2}$ foot. $3\frac{1}{2}$ feet can be written as $3 + \frac{1}{2}$ feet, or $1 + 1 + 1 + \frac{1}{2}$ feet, which is $12 + 12 + 12 + 6$ inches $= 42$ inches.

15. Anita's hair was 10 inches long in September.

$$\left(6 + \frac{1}{2}\right) + \left(3 + \frac{1}{2}\right) = 9 + 1 = 10$$

16 At 3:00, the angle will be $30 + 30 + 30 = 90$ degrees. At 4:00, the angle will be

$90+30=120$ degrees. At 5:00, the angle will be $120+30=150$ degrees. At 8:00, the angle will be $150+90=240$ degrees.

17. This is a rectangle, but the sides are not all the same length. So this is not a rhombus, and it is not a square. It is a quadrilateral because it had four sides. Is it a trapezoid? The answer is *maybe*. Some definitions say that a trapezoid has one set of parallel sides, so if a trapezoid has two sets of parallel sides, does it have one set of parallel sides? By one definition, it might have *exactly* one pair of parallel lines; by another it must have *at least* one pair of parallel lines, so it could indeed be classified as a trapezoid.

18. Draw an obtuse angle and turn it into a diamond with four equal sides. Make it narrow so that no angles look like right angles.

19. A four-sided figure that is both a rectangle and a rhombus is a square.

20. There are 6 lines of symmetry.

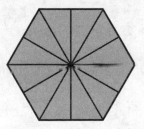

Chapter 9

1. The order of operations says to do all of the multiplication and division first. Working from left to right, multiply 1×5. The problem then becomes $11-5-4$. Now do all addition and subtraction operations, from left to right, to find the answer: 2. Neither Chad nor Megan are correct.

2. Evaluate the expression inside the parentheses first, and then simplify the expression $140 \div 10$, which is equal to 14 (B).

3. Simplify the expression inside the parentheses by multiplying 9 and 5. The expression simplifies to $52-45$, which equals 7 (A).

4. Marie first subtracts 9 from 17. This is represented by $(17-9)$. Then the result is multiplied by 4. This can be written as $(17-9) \times 4$. Notice that this exact translation is not one of the choices, but $4 \times (17-9)$ is one of the choices (C). Subtracting 9 from 17 is 8, 8 times 4 is 32, and the expression that matches Maries expression must evaluate to 32.

5. Correct matches:

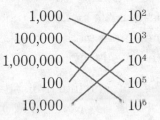

Each of the numbers in the right column has an exponent on 10, which indicates how many times the number 10 gets multiplied by itself. For example, $10^2=10 \times 10$, which is 100; $10^3=10 \times 10 \times 10 = 1,000$. The exponent on the 10 corresponds to the number of zeros following the 1.

6. Rounded numbers:

   A. The ones place is 8, so round up to the nearest 10, which is 10.

   B. Look at the hundredths place to round to the nearest tenth. It is a 4, so round down to 8.8.

   C. Look at the tenths place. It is a 4, so round down to 88.

   D. Look at the thousands place. It is 8, so round the hundredths place up to 88.46.

7. Multiples of 7:

   ☑ 287
   ☐ 400
   ☑ 308
   ☐ 456
   ☐ 421

There is a divisibility rule to test to see if a number is divisible by 7, but students are better off using other methods to see if the numbers can be evenly divided by 7.

8. The expression simplifies to $30 + 8 + 0.5 + 0.006$, which is 38.506 (D). Notice that there is a 6 in the thousandth place but no value given for the hundredths places, so it is 0. The thousandth place is a little beyond the fifth grade level.

9. The perimeter is 31 cm, and the area is 58.5 square cm.

The perimeter is found by adding the width on both sides and the top and the bottom along the length. $6\frac{1}{2} + 6\frac{1}{2} + 9 + 9 = 13 + 18 = 31$.

The area is found by multiplying the width times the length, $6\frac{1}{2} \times 9 = 58.5$. Using an area model or drawing a picture of the index card, you can see this is the sum of two areas, one is 6×9 and one is $\frac{1}{2} \times 9$. The sum of the two areas is $54 + 4.5 = 58.5$

10. One strategy is to create a number line marked every $\frac{1}{2}$ from 0 to 3 (or more), and comparing each number to $\frac{1}{2}$, 1, $1\frac{1}{2}$, 2, $2\frac{1}{2}$, 3. The only number that is less than one is $\frac{3}{5}$, so place it first. Next, see that there are two numbers less than 2: $1\frac{1}{2}$ and $\frac{5}{3}$. It may hard to compare $\frac{5}{3}$ to see if it is more than $1\frac{1}{2}$; try using an equivalent fraction, such as $\frac{10}{6}$. Can you tell if $\frac{10}{6}$ is more than $1\frac{1}{2}$? If you had two wholes divided into sixths, the $1\frac{1}{2}$ would be $\frac{9}{6}$, so $\frac{5}{3}$ is slightly bigger than $1\frac{1}{2}$. Next, compare $2\frac{1}{2}$ to $\frac{13}{5}$. If you represent $2\frac{1}{2}$ in one place and $\frac{13}{5}$ in another, you could remove two wholes from both places, and compare what is left. Subtracting 2 from $2\frac{1}{2}$ leaves $\frac{1}{2}$. Subtracting

2 wholes from $\frac{13}{5}$ subtracts $\frac{10}{5}$ and leaves $\frac{3}{5}$.

$\frac{3}{5} > \frac{1}{2}$ so $\frac{13}{5}$ is last.

$$\frac{3}{5} \quad 1\frac{1}{2} \quad \frac{5}{3} \quad 2\frac{1}{2} \quad \frac{13}{5}$$

11. This is a multistep problem with many ways to solve it. First, you can calculate how much lemonade they started with: 2 gallons is 256 ounces. Next, figure out how much they sold. They sold $25 \times 5\frac{1}{2}$ ounces. If they sold 5 ounces 25 times, that would be 125 ounces; if they sold $\frac{1}{2}$ an ounce 25 times that would be 12.5 ounces. So they sold $125 + 12.5 = 137.5$ ounces. $256.0 - 137.5 = 118.5$. They had 118.5 ounces of lemonade left over.

12. If they hand out one cone they use $2\frac{1}{4}$ ounces. If they hand out two cones they use $4\frac{1}{2}$ ounces. If they hand out four cones they use 9 ounces. This calculation not only measures the ice cream used in a whole number of ounces, but it is also a factor of 45. $45 \div 9 = 5$; they could hand out 5 groups of 4 cones, where each group of cones uses 9 ounces of ice cream. $4 \times 5 = 20$. They could make 20 ice cream cones.

13. The question supplies the size of each bottle in milliliters and asks for the answer in liters. There are 1000 milliliters in 1 liter, so convert

the size of each bottle to $\frac{1}{2}$ liter. There are 8 bottles of $\frac{1}{2}$ liter bottles. Multiply 8 and $\frac{1}{2}$ liter to find the answer in liters. $8 \times \frac{1}{2} = 4$ liters of water per 8-pack.

14. The sum of all of the whole units (pounds) is $4 + 1 + 1 = 6$ pounds. The sum of all of the ounces is $10 + 7 + 1 = 18$ ounces. Since 18 ounces is greater than 16 ounces (the number of ounces in 1 pound $= 16$), the 18 ounces should be rewritten as 1 pound 2 ounces. Add the 1 pound 2 ounces to the 6 whole number units of pounds, and the total weight of all items is 7 pounds 2 ounces.

15. To find the volume, multiply the length, width, and the height.

$$V = L \times W \times H$$
$$\text{Volume} = 9 \times 4 \times 2 = 72$$

16. To find the volume, multiply the width, the length, and the height. Here you are given the volume, the length, and the width, but not the height. Substitute the values that you know.

$$V = L \times W \times H$$
$$84 = 7 \times 4 \times H$$

This can be simplified by multiplying 7 and 4.

$$84 = 28 \times H$$

The value for the height that makes this true is 3. The height is 3 feet.

17.

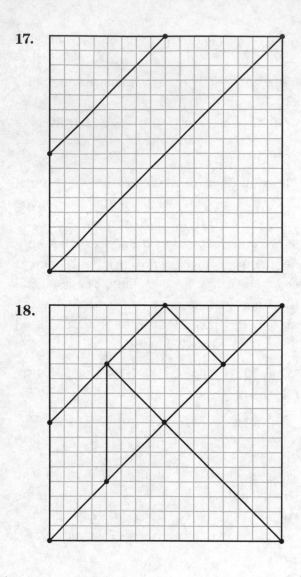

18.

19.

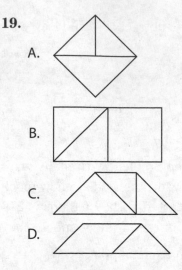

A.

B.

C.

D.

20. Each of the four answer choices are quadrilaterals. They all have four sides. Perpendicular lines form right angles. The only shape listed that must have right angles (and perpendicular lines) is a rectangle. All rectangles are parallelograms, but not all parallelograms are rectangles. A rhombus has four equal sides, and a square is a rhombus and a rectangle, but each rhombus isn't a square. A trapezoid could have perpendicular lines, but it could have no lines that are perpendicular. The only quadrilateral listed that must always have perpendicular lines is a rectangle.

APPENDIX C

Bibliography

Blackburn, Barbara R. *The Beginner's Guide to Understanding Rigor.* (2008), *www .barbarablackburnonline.com/rigor.*

Ginsburg, David. *Unlearning Learned Helplessness.* (Edweek.org, 2014), *http://blogs.edweek.org/teachers/ coach_gs_teaching_tips/2014/03/unlearning_learned_helplessness.*

Kendall, John; Ryan, Susan; Alpert, Alan; Richardson, Amy; Schwols, Amitra. *State Adoption of the Common Core State Standards: The 15 Percent Rule.* (Denver, CO: Mid-continent Research for Education and Learning, 2012)

Kilpatrick, Jeremy et al. eds; National Research Council. *Adding It Up: Helping Children Learn Math.* (Washington DC: National Academy Press, 2001)

Massachusetts Department of Elementary and Secondary Education. *Massachusetts Curriculum Frameworks for Mathematics Grades Pre-Kindergarten to 12: Incorporating the Common Core State Standards for Mathematics.* (Malden, MA: *www.doe.mass.edu, 2011*), *www.doe.mass.edu/frameworks/current*

Michigan State Government. *Pre-K Mathematics. www.michigan.gov/documents/mde/Math_DK_242336_7*

National Governors Association Center for Best Practices, Council of Chief State School Officers. *Common Core State Standards Mathematics.* (Washington DC: National Governors Association Center for Best Practices, Council of Chief State School Officers, 2010)

Probst, Gary K. *Overcoming or Preventing Math Anxiety That Causes Learned Helplessness In Mathematics.* (*www.learningassistance.com*, 2002), *www.learningassistance.com/2002/Apr02/Gary*

Recommended Resources

Arizona Department of Education. *Arizona's College and Career Ready Standards: Mathematics* www.azed.gov/azccrs/mathstandards

Bib Count School District, GA
schools.bibb.k12.ga.us

Brunswick County Schools, NC, Common Core Links
www.bcswan.net/education/components/ scrapbook/default.php?sectiondetailid=31611

Common Sense Media: Parent Concerns
www.commonsensemedia.org/parent-concerns

Common Core State Standards Initiative
www.corestandards.org

Curriculum Corner
www.thecurriculumcorner.com

Department of Public Instruction, Public Schools of North Carolina
www.dpi.state.nc.us

El Dorado Weather, Inc
www.eldoradocountyweather.com

English Plus: Abbreviations of Units of Measurement
http://englishplus.com/grammar/00000058.htm

Elizabethtown Independent Schools, KY
www.etown.k12.ky.us

ESPN
espn.go.com

Georgia Standards, Georgia Department of Education
www.georgiastandards.org

Howard County Public Schools, MD
www.hcpss.org

Illustrative Mathematics
www.illustrativemathematics.org

IXL Learning
www.ixl.com

K–12 Center at Educational Testing Services
www.k12center.org

Math Goodies Aligned with the Common Core Standards for Mathematics, Grade 5
www.mathgoodies.com/standards/alignments/ grade5.html

HWMath.net
http://hwmath.net

National Council of Teachers of Mathematics, *www .nctm.org*; see also their hosted web pages for the Math Common Core Coalition, *www.nctm. org/standards/mathcommoncore*

New York State P–12 Common Core Learning Standards for Mathematics
www.engageny.org/resource/new-york-state-p-12-common-core-learning-standards-for-mathematics

Oregon Department of Education, Common Core State Standards Toolkit
www.ode.state.or.us/search/page/?id=3511

Partnership for 21st Century Skills
www.p21.org

Smarter Balanced Assessment Consortium
www.smarterbalanced.org

Tangipahoa Parish School System, Common Core State Standards
www.tangischools.org/Page/18155

U.S. Department of Education Institute of Education Sciences, National Center for Education Statistics
http://nces.ed.gov

U.S. Rice Producers Association
www.riceromp.com

Word Hippo
www.wordhippo.com

Index